Dr Jordan Nguyen is a biomedical engineer, inventor, TV documentary creator and presenter. He is founder of Psykinetic – an innovative social business that harnesses cutting-edge advancements, including biomedical, robotics, cloud, extended-reality technology and AI, to create life-changing, empowering and inclusive technologies aimed at improving quality of life for people with disability and the elderly. He has featured at TEDx, spoken to hundreds of thousands of audience members across numerous domestic and international events, has his own production house, Time-Captain Studios, and is an award-winning host of several documentaries, including *Becoming Superhuman* and *Meet the Avatars* for ABC Catalyst and *Smart China, Living on the Roof of the World* and *Vietnam: Connecting East Africa* for Discovery Channel. *A Human's Guide to the Future* is Dr Jordan's first book.

A
HUMAN'S
GUIDE TO
THE FUTURE

HOW HUMANITY CAN THRIVE THROUGH
OUR EVER-EVOLVING TECHNOLOGY

DR JORDAN NGUYEN

MACMILLAN
Pan Macmillan Australia

First published 2020 in Macmillan by Pan Macmillan Australia Pty Ltd
1 Market Street, Sydney, New South Wales, Australia, 2000

Cataloguing-in-Publication entry is available
from the National Library of Australia
http://catalogue.nla.gov.au

Typeset in 12/18 pt Fairfield LH by Midland Typesetters
Printed by IVE
Illustrations by Lesley Nguyen and Dr Jordan Nguyen;
illustration on page 132 by Justin Currie; technical layout for illustrations
on pages 117, 167, 187 and 302 by Hannah Schubert.

The author and the publisher have made every effort to contact copyright
holders for material used in this book. Any person or organisation that
may have been overlooked should contact the publisher.

MIX
Paper from
responsible sources
FSC
www.fsc.org   FSC® C018183

To my loving parents, grandparents and family, who have been a consistent light in my life, my inspiration. I love you more than words can describe.

To my future family, my children I have not yet had. I hope we did everything we could and that the world is a good place – this is also for you.

CONTENTS

Technology is just a tool.
What's fascinating is how we choose to use this tool,
for tomorrow may be bright and rich with experience,
but that's up to us.
It should never be about the technology itself.
It's all about the human odyssey,
and what we learn about ourselves along the way.

INTRODUCTION

LIFE MAY BE CLOSER TO FANTASY THAN WE THINK

Our world is advancing at a pace we've never seen before, and it can be overwhelming. Think of how much you've already outsourced your cognitive processes to your smart devices and sometimes feel like part of you is missing if they're out of battery or left at home. Your dependence on GPS to find your way along that route you always take but never seem to remember. Those voice-activated assistants that always make us question if they're really *listening*. Facial-recognition ID that sometimes likes to irritate us mildly by not recognising us when we're overly tired and looking a little rough. And, of course, heart-rate monitors, blood-pressure sensors, those bits-for-tracking-the-fit and all the other techy stuff we find ourselves hurled into love–hate relationships with. Honestly though, you ain't seen nothing yet!

The thing is,

We're living at the fastest rate of change
the world has ever seen.
But it may also be the slowest
we'll ever see again.

The rate may seem to slow here and there when events occur with global impact, but they barely affect advancement when considering the big picture. Living in these times of rapid change means we should consider how technology will play a defining role in the world of tomorrow. In all my travels and among the many people I've met, what I've found is that this both excites and creates a sense of fear. Do we try to resist these ever-evolving changes coming our way or do we learn to embrace them? How *can* we embrace them? Will they lead us to a better future or will we fall into turmoil and chaos? Whether we like it or not, we're having a massive impact on our earth. As the human population expands, so too does our consumption of natural resources. It is within our capabilities to harness technology to benefit the planet: to better model, understand, anticipate and steer our global changes and challenges. All major advancements come with new sets of ethical issues that our world needs to work through. I believe the present is an ideal time not only to explore the positive sides of these shifts, but also to activate our individual and collective potential.

Through my broad range of adventures I've met people from all walks of life, and I've become increasingly certain that collectively we can create a better future – absolutely.

The way I see it, our only way forward is to embrace and guide these technological revolutions and work together to ensure they are used for the good of humanity and our planet. I've loved technology for almost as long as I can remember. I've always been fascinated by visions of the future offered by science fiction – whether they be exhilarating or cautionary – but for me it's never just about the technology itself. It's about our dreams, humanity, purpose and finding the positive global impact we can achieve through bringing our imagination into reality.

LET ME INTRODUCE MYSELF...

I'm Dr Jordan Nguyen, an engineer, inventor, documentary presenter, speaker and entrepreneur. My imagination runs wild and I chase the ability to turn dreams into reality. I want to show you the latest visions for the future and share my adventures as I learn that nothing is impossible, find great sources of inspiration to tackle big human-fuelled ambitions, and assemble super teams to bring our ideas to life.

I'm the founder of Psykinetic, a Sydney-based social business where we use the latest biomedical, bionic, robotic, artificial intelligence and extended-reality technologies to create inclusive and empowering solutions aimed at improving independence and quality of life for those with disabilities, our elders and beyond. I've developed many projects over the years including a robotic tour guide at university; avatars to preserve the memories of loved ones; and a smart wheelchair that could make its own decisions in navigation and, coupled with unique biomedical technology developed during my PhD studies, could even be controlled by the power of the mind. I've also worked on automated software systems for medical equipment, life-support systems, 3D-printed motor prosthetics, smart-home systems, virtual-reality rehabilitation, and eye-controlled systems for communication, gaming, transportation and music creation.

On the media side I founded Time-Captain Studios primarily to create documentary films that explore science and technology working towards a better future. I travel the world, usually with documentary crews as a presenter and journalist (having worked with production groups including ABC Catalyst, The Feds, Meridian Line Films, CICC, Matavia and Beach House Pictures), to witness first-hand the effects of science and technology across the globe. I've been through China to see how their mindset has been shifting and to experience the fast pace of innovation from their startups to their groundbreaking scientific advancements. I've been around the US to see cutting-edge research and meet some of the great minds behind many amazing technologies like applying virtual reality to psychological rehabilitation and advancements in biomedical technology, to name a few. I've been to Vietnam to discover how its focus on telecommunications infrastructure helped its people pull themselves up from being one of the poorest nations after the Vietnam War to thriving in the top twenty fastest growing economies in the world. And how, in a great tale of economic kinship, they've partnered with East Africa to help them emulate the same lessons, working with them to build their own telecommunications infrastructure to unbelievable levels within short time frames. I've been through Tibet to see how humans have come to thrive at high altitude and make unique uses of technology. I've gone into the tunnels with the rail workers and engineers who are building the most challenging trainline in the world; visited large-scale stations there that can measure cosmic rays from galactic events like supernovas; climbed the declining Tibetan glaciers with scientists to take samples for predictive modelling on how long they may last; and trekked to some of the most remote regions of Tibet, where technology has even reached villages rarely visited by foreigners.

I've been fortunate to meet world-changing people through my professional speaking. In 2015 I was joined on stage by Nolan Bushnell, founder of Atari and the first person the late co-founder of Apple Steve Jobs ever worked for. In the same year I shared the stage in panel with artist, designer and director Ron Cobb, creator of the iconic time machine DeLorean for the *Back to the Future* films. In 2016 I travelled with Steve Wozniak, co-founder of Apple, and was master of ceremonies for his Australian speaking tour with Think Inc. In 2018 I even had the honour of being master of ceremonies for 'An Evening with President Barack Obama' on his visit to Sydney.

I bring so much of my learning back to Australia and continue to create with purpose, keeping up with what's happening outside the bubble of our own country. I feel like I've racked up the experience of a few lifetimes and yet I'm still only a few steps into this journey.

The most important point I should make – I'm no genius. In fact, I struggled throughout school to be even an average student and I almost dropped out of university soon after starting. Without realising it, I was waiting for a spark of inspiration to hit me and become my fuel. That inspiration turned out to begin with human connection and spread through to imaginative endeavour, humanity and a desire to collectively build a better future.

I'm a thinker with endless curiosity, an engineer who has moved through electronics, mechatronics software and biomedical design and development, a future dreamer, an inventor and innovator, an author, a TV documentary creator and presenter, a tennis player, an adventure-seeking, people-person, car-singing, always-dancing, pasta-loving, vision-pursuing, family-oriented, stargazing, art-admiring, knowledge-consuming, problem-solving, mostly-laughing, sometimes-curled-up-crying, life-loving, crazy-haired, empathy-filled giant kid . . . But above all, I consider myself to be an everyday human who has taken big leaps to live my dreams and visions. I will continue to work towards making the changes I want to see in the world, and I will challenge you to do the same.

WHY THIS BOOK?

I wrote this book because in many ways I'm probably a lot like you – I lived a reality in which I knew there was more. I pursued the vague dream that plagued me, the hope that my life could make a difference, that my existence was to be earnt, not just used up. My mind has been opened and stretched over and over, but I still feel like I'm only a few steps into this audacious journey, because the world is a big place to improve.

People often wonder if I'm scared of technology and where it could be headed. The truth is I am concerned about where it could end up, but I don't let this concern turn into fear that rules me. The energy of fear can be faced and converted – transformed into motivation and optimism to shape positive change – and that takes courage. But I cannot do this alone. We need to connect more, share more, do more, and together write our way into that unwritten future.

I find that the more a mind is opened to the reality of emerging possibilities, the more it's susceptible to believing in what could be made possible next. When you know about what's going on, you can have a say in steering the things to come. So this book is made for those who are curious about how fantasy and reality collide, how technology and humanity intertwine and who want to join me on a journey I've come to realise is too wondrous to not share through this medium. Let's learn from what has been, let's gain a brief understanding of what is, and let's tilt our imagination towards our future, which seems more uncertain (and hopefully more exciting) than ever before.

THE SUPERHUMAN ERA

Many humans are scrambling to understand how they and the world around them will be affected as we move through a new era. Some call it the Fourth Industrial Revolution. A technological revolution. A massively pivotal moment in human history that will fundamentally change the way we live, work and relate to each other. It will chart the course of our future.

In each industrial revolution new inventions have changed the fabric of society. The First Industrial Revolution began in Britain in the mid-1700s, underpinned by the invention of the steam engine, which enabled new manufacturing processes and led to the creation of factories. The Second Industrial Revolution came about a century later, with the emergence of mass production in new industries like steel, oil and electricity. Notable inventions of this era include the telephone, the internal combustion engine and the light bulb. From the 1960s, the Third Industrial Revolution arrived. Also known as the Digital Revolution, this was marked by the invention of semiconductors and silicon chips, personal computers and the internet.

When people talk about the Fourth Industrial Revolution, they're usually referring to the merging of the physical, digital and biological. In other words:

Get your undies on the outside of your pants
and slap on your best cape,
because we're hurtling towards a new era.
A *superhuman era*!

Unlike the major advancements of the past, we're now facing a vast range of disruptive technologies that could all in their own right change the world. Like all those that came before, each of these can change the way we live and work, can impact the behaviour of society, will bring up new ethical issues and dark sides, and will also give rise to new positive opportunities to better our world. But the biggest impacts will come from the creative *convergence* of a number of these to create super solutions of awesomeness. They will make the impossible possible.

Let's work together,

towards acting as a self-aware, intelligent species,
towards thriving and flourishing,
towards exploring the stars,
towards imagination,
towards equality,
towards peace,
towards survival,
towards a better world,
towards creating our next evolution,
towards moving beyond our present form of human,

and instead . . . becoming superhuman.

I will guide you through what I believe to be foundational technologies as we move into this new era. These have resulted from incredible events and advancements from the past, and are now all branding their mark on the world. This vast range of rising technologies includes robotics, artificial intelligence, big data, cloud computing and cloud services, bionics and biomedical technology, extended reality (such as virtual, mixed and augmented reality), the internet of things, 3D printing, nanotechnology, genetics and genomics research, blockchain technology, cryptocurrencies, renewable-energy technology, advanced sensory systems, quantum computing and much, much more.

Although these are the technological topics, you'll find that I always approach them from the purpose and human-impact side, harnessing these tools as I go, rather than it being about the technology. The chapter titles in this book may seem a little counterintuitive, but they are all a play on my philosophy that if we must embrace technological advancement, then we should also humanise it. Throughout these topics, we'll delve into the past, present and where I envisage the future to be moving. I'm always dreaming and imagining the future. My mind is firmly planted there while making the most of every moment in the present and never taking anything for granted.

This book is my guide to the future, with ideas of where advancing technology will take us. I don't just dream ahead and talk about tech, I harness it to make audacious ideas a reality and create the things that I've envisaged and that I believe need to be made. I've cultivated my life so I can chase big ambitions and take action, so all of these topics will be interlaced with my adventures as I've designed technologies that have made some big and sometimes seemingly impossible dreams a reality – building on the past, 'standing on the shoulders of giants', utilising the latest technologies and harnessing the power of collaboration with many amazing people I've met along the way.

Over the years, I've built up a unique set of skills – from inspiration to action – to become the architect of solutions to difficult challenges; to design, create and build what is needed; to rally energetic teams and rollercoaster our way through innumerable hurdles and seemingly insurmountable roadblocks; and to experience mind-altering, reality-expanding moments when a new (and often emotional) idea is finally realised.

As overwhelming as these topics might seem, trust me, it's not too difficult to dabble. We just need a bit of fun, some very human analogies and challenges worth solving for the betterment of life and the planet. Even a basic understanding can allow us all to have a say on where the future is headed. I'll show you what's been made possible and what I've seen. I'll take you into the massively ambitious, purpose-driven projects I've worked on that utilise these technologies. This book will delve into some seriously exciting stuff, involving robot friends, merging with artificial intelligence, hacking our senses, creating superhuman abilities, preserving loved ones through avatars, exploring what it means to be human, bending reality and fabricating virtual universes.

So, my friend, ready that cape you're now wearing and take my virtual hand as I guide you through my mind and our ever-evolving reality . . . because adventures await!

We could be heroes,
like those who roamed before.
We could leave an eternal legacy.
A better future starts with us.
Change starts with us.

PART I
BEFRIEND A ROBOT

You may learn more about
what it means to be human

1

Δ ROBOT WΔS ONE OF MY FIRST FRIENDS

A LIFELONG FASCINATION BEGINS

This journey starts with my first love of advanced technology . . . Robots! My fascination with robotics originated from two major influences as a child. The first was Johnny Five from the 1986 science fiction film *Short Circuit*, and the second was a real-life robot I knew from a young age – an industrial robot my engineer father was researching and developing at the University of Technology, Sydney (UTS). The movie made a lasting impression on me, and Dad's work had so much crossover it made me think what I was seeing in the movies could be real.

Short Circuit's plot revolves around the last in a line of five experimental military robots, aptly named Number 5 – no prizes for guessing the names of the first four. They're slim, tall metal robots with tank-like caterpillar wheels, incredible dexterity in the movement of eyes and eyebrows, and was a clear influence behind the starring robot's design in the animated Pixar film *WALL-E* a few decades later. Following a demonstration of the five robots' capabilities to the military stakeholders, an electrical storm rolls in. Just as the small robot fleet starts moving for shelter, Number 5 is struck by lightning. His circuits are scrambled, he loses all direction and ends up among the rubbish bins being transported out of the compound. A complete mess lost in the outside world, Number 5 reboots his system and starts the search for information to help him find his bearings and gather data about his environment. He frequently says 'Need input' out loud – feeding his hunger by reading books, watching TV and observing the world around him. He slowly gains understanding and builds an empathetic personality. He even starts developing sentience – the ability to feel, perceive or experience subjectively. He decides he is *alive*!

Throughout the film, whenever he got shot or blown up, I would run out screaming to Mum. Bear in mind here I was only five years old. I remember Mum once responding, 'He was okay last time wasn't he? Go watch the rest to see what happens. I'm sure he'll be okay again this time.' She was right. He was fine.

But why was I crying? He's only a robot, in a movie. I guess the story just got to me in a way that made me feel as if, as he himself puts it, 'Number 5 is alive'. This meant I didn't want to see him get hurt. It affected me at that young age because it was already teaching me that life is precious. After an accident where Number 5 jumps on a grasshopper and kills it, he tries to comprehend the permanence of the event, and links the concept of death with being disassembled. He soon learns this is exactly what the military want to do with him when they catch what they think is just a malfunctioning, rogue robot. He doesn't want to be disassembled, yelling, 'No disassemble' at the prospect of death. I didn't want that for him either. Thinking about it now, I can see what a brilliant way it was to teach kids about the concept of death. He later takes his independence in thought and experience a step further and chooses to rename himself Johnny Five.

It makes you wonder how the next generations will feel as they grow up with robots as normal parts of society. Robots that can simulate and exhibit these human traits. As they start to grow *curious* – however artificial that curiosity might be – will that make us believe that they are sentient? Will we learn to empathise with robots? Will we treat them like pets, or wild animals, or maybe even afford them close-to-human status?

The realms of sci-fi and technology have provided lifelong fascination for me. We didn't have too many toys growing up so we relied heavily upon our imaginations. By we, I mean me and my siblings. We are the product of two beautiful, very different parents. Dad was born in Vietnam and came to Australia as a teenager at the start of the 1970s on a scholarship scheme known as the Colombo Plan to study electrical engineering. During these studies, his family in the south of Vietnam lost everything as the war spread, and he decided to stay and work in Australia after completing his bachelor's degree and PhD. He has a truly brilliant mind, an incredible knack for technological foresight and the engineering skills to boot. Mum is Australian-Scottish and a wonderful, social, compassionate person who, among many other skills, has been an artist her whole life and even helped me with the illustrations in this book. She provides me endless inspiration as to the beautiful effect selfless love and empathy can have on the people around us – it's simply her natural aura.

After I was born, my parents planned on having just one more child. But just three days shy of three and a half years after my birth, they were landed with three at once. A lucky miracle! They had triplets and I suddenly had a sister and two identical twin brothers: Zohara, Alexander and Tristan. From then on, I had companions to share in the trouble I got myself into. Until I was seven years old, we grew up in the western Sydney suburb of Merrylands.

Ah, Merrylands, a land of adventure. We would make the best of our imaginations and often created dreamlands to which we all made contributions. Anything and everything could be a prop for these games, and we were heroes. They were simple times, but we found great excitement in running around the neighbourhood. Imagining. Exploring.

Years on, my brothers are both medical doctors and my sister is a speech pathologist. I couldn't be more proud of my siblings and the outgoing, kind people they have become. I'm often asked how intellectual our family dinner conversations are. Honestly, we often talk about many everyday things, mostly about music and movies, the latest thing my little niece Elora has learnt and a lot of random stuff. We have the odd argument. We do karaoke. We play board games like *The Settlers of Catan*. And we just generally appreciate being in each other's company . . . but not for offensively long periods of time – we've learnt the value of balance. To this day, I still feel that none of us has lost our childhood imagination.

Heavily influenced by the books we read and films we watched throughout our lives, I, for example, let the heroes, the superhuman abilities and the endless realms of possibility stay in my mind. I've also never forgotten the dystopic depictions of what the world could potentially become, especially as I've experienced events that blur the lines between sci-fi and reality, some of which I'll share with you through this book. Perhaps unlike other people of my vintage, I realised from a young age that technologies like robots and artificial intelligence (AI), the themes of *The Terminator*, actually existed in the real world, and I started treating these scary movies as warnings of what the future might bring. Over the years, we've seen more and more imagined technologies stripped out of the realms of science fiction and into our everyday lives.

Whether or not you're a lover of science fiction, it is largely to thank for where we are today. Sci-fi doesn't necessarily predict. What it does do is it gives us ideas for the future. When those ideas seemingly become possible given the current tools available, people tend to make them real. In some cases, advanced technology, such as AI, has been around for many decades, but it takes a big event (like an AI winning chess or Go against the best the human race has to offer) for it to spread around the world and scare the bejesus out of some, and entice the curiosity of others. But why does a machine winning a board game incite fear of our entire civilisation being wiped out? Simple. Because we suddenly link the technology to sci-fi like *The Terminator* and quickly imagine the entire story possibly becoming a reality too. In situations like this, we need to keep a cool head, sort fact from fiction and remember that we are in control.

In addition to what I was seeing in the movies, my father was working on a real-life industrial robot during my childhood years. I began noticing that some Duplo blocks we had growing up were disappearing, and discovered that Dad was taking them to work with him. At this stage all I knew about his work was that it involved computers and the odd gadget he brought home. Oh, and that he could type fast. Lightning fast. I used to love just watching him at the computer and listening to the clickety-clack sounds of him speed-typing away.

When I was six years old, the day finally came for me to see what Dad did every day he left the house. I went into work with him, to the giant tower building of UTS. Once voted Sydney's ugliest building, it's what's on the inside that counts, and this tower was full of wonder. It really opened my eyes.

I can picture the memory as if it's happening right now. My father walks me through to the lab where he has a large industrial robot. It has my Duplo blocks and suddenly the penny drops. I'm sharing my toys with a robot. A *real* robot. I thought these things only existed in the movies. From the mid-1980s my dad and his research students have been repurposing a UMI brand RTX model robot arm to give it new abilities. It's basically a big pillar with a large claw sticking out, all grey and beige. It has a camera in its claw, a camera on the top of it facing down, a large computer next to it displaying the top camera's view on the monitor and a conveyor belt running across the front of it. It's gigantic and incredible and mesmerising – and *my* dad is working on it.

Time to see how it works. Dad types a few commands into the computer and places my Duplo blocks on the conveyor belt as it starts to move slowly. The robot hand gives a few quick twists around the wrist then heads over to hover directly above a block. It *watches* the block with the camera in the claw, and follows it, as if *fascinated*. With a few sudden movements it descends with an open claw, reminiscent of a skill-tester machine reaching for a plush toy prize. It picks up the block, swiftly takes it to another location on the belt and gently puts it back down.

Now this might seem like a simple task, but there are many things to program into the robot to achieve this. It has to recognise the shape of the block, the size of it, how it's going to get its claw around it, the velocity it's moving at, how quickly the claw can move, where the intercept point will be – and then execute the manoeuvre. However, instead of programming all this directly, Dad built in artificial intelligence, an ability to learn for itself in real time (learning and updating during the actual time real events are taking place). It learns by trial and error, attempting to pick up the block, working out where it went wrong, updating and trying again. It's not afraid to make mistakes. It's not afraid to fail.

That's just it though. Why would we ever build a fear of failure into robots? We don't, which is why they get so good at what they do. Imagine the alternative: if robots were like humans tend to be, and weren't willing to try certain tasks because they were afraid they might stuff it up. A few more steps on and we'd end up with Marvin the Paranoid Android, the 'manically depressed' robot from *The Hitchhiker's Guide to the Galaxy*. It's something we can reflect on for ourselves too. Something we're beginning to teach the next generation these days. Do not let fear of failure take over. It is okay to make mistakes. Because it's through these that we learn, often more than through our successes. If we let fear get in the way, we might never try in the first place.

And so the RTX is ready to tackle its next challenge. It knows how to move blocks around, but can it adapt to a Duplo-block horse? To a human, even a toddler, this is a simple shape. To a robot, well . . . Dad places the Duplo horse on the moving conveyor belt. The RTX activates, moves the claw above the horse, hovers, *watches* and starts to follow. This time it isn't making a move for the block, it just keeps following, keeping a 30 centimetre height above the horse. It's running out of track and yet it still follows. Suddenly the RTX starts shuddering, loudly. Dad makes a dash for the computer to shut it down but before he can, the system crashes. The monitor goes blank and the RTX goes still. 'Huh. Strange,' Dad says, pondering what just went wrong. He reboots the computer. 'Let's try again.' The horse moves along the conveyor belt, the RTX activates, follows, as Dad, his students and I all watch silently. Again it follows, further than any previous attempt. It starts shuddering but continues on. It makes a move! Straight down towards the horse, and . . .

BAM!

It misses, crashing through the conveyor belt and breaking itself on the metal support bars underneath. Dad lunges for the computer and shuts it down. 'Agh! Is he all right, Dad?' Hold up. Did I just say 'he'? I guess he naturally does seem male. Maybe it's the bulkiness or the clumsiness, I'm not entirely sure, but that's his persona. RTX definitely seems like a *he* to me. Interesting.

A few minutes later, with the help of some good ol' electrical tape (an engineer's best friend), RTX's broken claw is fixed and ready to rock. Back to it. Dad is determined to see this robot master the task. Miss after miss, RTX shudders less and less. He improves to the point of being tantalisingly close to grabbing hold of this darn horse. And then . . . Success! RTX manoeuvres his claw around the horse as he moves past, then carries it to another area. From then on, he no longer seems to have the same trouble with this shape.

For many days following, the events whirl around on repeat in my mind. It was just so cool to witness the RTX trying over and over, and developing a personality as he went. A resilient personality. My father's work is surreal and I cannot wait to see it again. Talking about it at school over the next few weeks, I confirm my thoughts on the matter – this is not normal. Other kids' parents don't work on robots, and some friends respond with doubt that mine even does. No matter. I know what happened.

My next visit to RTX arrives a few weeks later. Excitement! Now he's playing board games. He's been programmed with the rules, but needs to learn strategy through experience. I get the first go at playing him in noughts-and-crosses. Every kid knows how to play this game anywhere; as kids we'd draw it with pencil on paper, or with chalk on the pavement, or with our fingers on fogged-up school-bus windows. Now RTX is having a go. Here, the game is played on a flat grid with marked cylindrical felt-covered wooden pieces that can be identified from above. These make it easy for RTX to see the pieces with his cameras and to move them around. This is a fresh program for RTX and he has no prior experience of it, but I don't know this.

Game on. RTX goes first. I'm just so taken with the way he sits as if he's thinking and, when ready, twists his claw a little and moves to pick up one of his pieces off the edge of the play area, moving it into place on the board. I love the *vvt . . . vvvv . . . vvt* sounds of the motors as they facilitate the heavier actions of the arm articulation and lighter actions of opening and closing of the claw. He's zippy in between picking up the pieces and placing them down again, at which times his motion slows. Smoothly. Carefully. *Intentionally*.

He takes the right-middle edge square. I instantly go for the centre square wondering why he didn't make that choice instead of an edge. His second move, taking the left-middle edge, seems ridiculous. I'm quickly realising this robot either doesn't have a clue about the strategy involved, or he has some genius moves I cannot yet comprehend. I claim the top-left position, expecting him to block me from my next winning move by grabbing the bottom-right corner. But he doesn't. He takes the bottom-left corner. I stare at it for a moment – did I miss something? Nope. It really was another silly move. I place the finishing move to take the bottom right and win with a diagonal three-in-a-row. From his programming, RTX knows I've won.

Strange. This is not how I've come to know robots. I thought they were smart. This is a simple game yet his moves were all over the place. So we try again. It's my go, as we take turns to make the first move of each round. I take the centre square. Moments later, I win. Another round. Again he goes for a set of strange moves. I win. For six rounds straight I beat RTX. But then, on the seventh round, he starts by taking the centre square. Is he using my moves against me? Possibly. This time we draw. He continues to use the tactics that won me the first six. We draw again, and again, which to be fair is what happens when

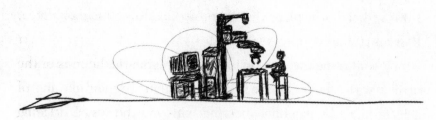

both people know this very basic game. But just like before, with his attempts to pick up the horse, what I'm witnessing is RTX *learning* for himself.

We move from noughts-and-crosses to *Connect 4*, and then chess – which I love, since Dad has taught me from a young age. RTX and I continue through these various games and with each it takes significantly more rounds for him to start playing as well as he did the previous game – due to the increasing complexity of the games and of RTX's learning curves.

During chess, the activities take a new turn. This is the first game in which we can both remove pieces from play, but only I've taken his pieces thus far. After eight flawless checkmate rounds as RTX flails about as a complete novice, he finally looks poised to strike. I can tell this because he's hovering over one of my pieces instead of his own – a pawn that he can take with one of his knights if he chooses. He makes the move, and while doing so he lifts said pawn off the board, transfers it out of bounds and doesn't place my pawn down in the careful manner he has with every other piece up until this moment. Nope. Not at all. Instead, with his claw still 30 centimetres above the bench, RTX just opens it and dumps my pawn, which bounces right off and to the floor. Shocked and unsure why he did that I quickly hop off my stool to pick it up. But as I rise up, what do I see? RTX swiftly moves his knight into the position he took,

slowing down as he places it. Smoothly. Carefully. *Intentionally* . . .
Bad robot, bad!

He's taunting me. I've no idea who programmed him this way
or if it was a fault. Doesn't matter though. I kinda like it. It's
a little like he's playfully messing with me the way I do with
my siblings. RTX has quite a character. I can see him learning,
and the more we interact, his personality seems to grow too.
I mean, it's like he *has* a personality.

But he can't. He's just a robot. I mean . . . *it's* just a robot.
Right?

Why would we ever build
a fear of failure into robots?
We don't.
That's why they get so good at what they do.
It's an analogy we might like to reflect on,
even as humans.

2

THE FOUR LAWS OF ROBOTICS

WHAT WAS ASIMOV ON
ABOUT ANYWAY?

Science fiction author Isaac Asimov famously devised the Three Laws of Robotics, first introduced in his 1942 short story 'Runaround'. Alongside coining the term *robotics*, Asimov's laws formed an underlying basis for his ongoing robotics-based fiction. Initially intended as a safety feature, they cannot be bypassed by the robots within these stories. These have over the years been altered and elaborated upon, but the basic laws are as follows:

1. **First Law**: A robot may not injure a human being or, through inaction, allow a human being to come to harm.
2. **Second Law**: A robot must obey the orders given it by human beings except where such orders would conflict with the First Law.

3. **Third Law**: A robot must protect its own existence as long as such protection does not conflict with the First or Second Laws.

In later fiction, Asimov even added another preceding law, a Zeroth Law, to account for the robots responsible for governing whole planets and human civilisations, and interacting in human societies to a far greater extent than previously written. This additional law is:

0. **Zeroth Law**: A robot may not harm humanity, or, by inaction, allow humanity to come to harm.

This basic framework, made to ensure robots served humanity and not the other way around, was very clever for its time. Asimov imagined a future world in which robots were commonplace, and created these laws before rudimentary forms of artificial intelligence – or even the first PC – existed. They did, however, also embody one fundamental misconception: that robotics always come packaged with AI, and vice versa.

Before we go any further, we need to sort out what constitutes a robot and what the difference is between a robot and AI. To make it even more fun, definitions vary wildly between experts and sources. But to me, a robot is as an autonomous, semi-autonomous, or remote-controlled device or machine that perform tasks otherwise done by people, one that is able to sense, compute and act.

I've come to think of it as five common bits that make up a robot (although a robot doesn't necessarily need all of these, they will vary broadly from robot to robot).

A HUMAN'S GUIDE TO FIVE COMMON ROBOT BITS (RB-5)

1. **The *sensing* bits**

   Sensors that absorb information about the environment – these are the *inputs*

2. **The *brainy* bits**

   Computation and control systems – whether *intelligent* or not

3. **The *doing* bits**

   Various movers, actuators, manipulators, propulsion systems, end effectors – these are the *outputs*

4. **The *communicating* bits**

   From simple speakers and microphones to wireless communication systems – these help the robot communicate and be communicated with

5. **The *power* bits**

   Often in the form of batteries, these systems provide the *energy* so all other bits can do their thing.

Figure 1: Five common robot bits (RB-5)

Their form (how they look), function (what things they can do) and dexterity (how fancily they do all those things and how difficult those things are), change from robot to robot, so starting with a solid purpose is a good way to figure out how all the rest of these building blocks will play out.

Another couple of quick definitions before we go on. An android is an artificial human, a robot made to look as human as possible – nigh indistinguishable; while a humanoid is simply a machine whose shape is inspired by the human form – you won't be tricked, it's clearly a robot.

As for the difference between a robot and AI, let's think of the robot as the *body* and AI as the *mind*. Sure, they may often go hand in hand, and the presence of AI can enhance a robot's capabilities, potentially to unfathomable heights. But robots can be and have been very useful without AI, and likewise AI has a vast range of uses without the physical manifestations of robotic bodies.

Asimov's laws are still often referred to today, as robotics and AI agendas mature and more of these discussions pervade our day-to-day lives. While Asimov's laws clearly combine both AI and robotics, and assume robots have some form of *understanding* of the world around them, some of the robots discussed in this part of the book have AI and some do not.

The etymology of the word 'robot' dates back to the early twentieth century, when Polish playwright Karel Čapek presented a uniquely futuristic play called *R.U.R.*, or *Rossum's Universal Robots*. The word 'robot' was based on the Slavic-language term *robota*, which basically translates to 'forced labour'. In this play, robots were created to serve humans, but over time adopted their own ideologies and attempted to overtake them through extermination . . . Bad homicidal robots, bad! Our understanding of the term 'robot' has changed since then, and more recent discussions edge closer to these entities even developing into human *companions*, which is clearly a big promotion from *forced labour*. Where a great level of fear lies is in the idea that they may one day, through sentient-like AI, promote themselves to *human overlords*, as has been the case in many sci-fi stories since *R.U.R.* – in which case we're all just doomed . . . dooooooooooomed!

Really, though, it's more likely we'll make friends with our emerging non-biological counterparts than be destroyed by them. But that's also partly due to how we as humans have evolved what we want and need from these entities. Our understanding of robots has changed over time. These days the key defining factor tends to be *automation*. They are mostly seen as machines of some form that can provide assistance or even take over physical tasks – of a heavy, strenuous, dangerous, menial or repetitive nature. In more recent times, however, they're not just automating these physical tasks for efficiency. They're also starting to be used in roles where there's a deficit of human physical entities, such as in aged care, when people can't always be present to keep residents company.

Early signs of people thinking about creating machines that exhibit forms of human characteristics or intellect date as far back as Ancient Greece, with their ideas of robotic men. Greek philosopher Aristotle wrote, 'If every tool, when ordered, or even of its own accord, could do the work that befits it . . . then there would be no need either of apprentices for the master workers or of slaves for the lords.' In the late fifteenth century, Leonardo da Vinci designed the first recorded 'humanoid robot' – a robot that takes on human form – with his complex mechanical Automaton Knight, operated by pulleys to give it the ability to sit up and stand, move its arms independently and even raise its visor. Although no one knows if this was ever actually built, da Vinci's technical design notes (whose style I just love) have been used in recent years to build the knight and it functions completely as Leonardo intended. An astounding feat of technical brilliance for the time.

In examining the history of robots, we need to look at both fictional and real-life approaches, as each has influenced the other. In 1881, Italian author Carlo Collodi introduced *Pinocchio*, a story about a wooden marionette, made by a woodcarver named Geppetto, who has his wish granted by a fairy of becoming a real boy. It wasn't exactly a robot story, but the theme of a human-made creation having the desire for life has woven its way through many robot-centred stories ever since. Back in the real world, at Madison Square Garden in New York in 1898, Nikola Tesla demonstrated a new invention he called a 'teleautomaton' – a radio-controlled boat. He was so far ahead of his time that his audience believed it to be a trick. Years would pass before this technology became more commonly used, such as in remote-controlled weapons in World War I during the 1910s. Fictional robots appeared in silent film in 1927 with the release of *Metropolis*, based on a novel from a few years earlier. Set in an urban dystopic future the film featured an iconic humanoid female robot, Maschinenmensch (German for 'machine-human'), who takes the form of a human woman, Maria, in order to destroy a labour movement. Again with the destroying!

Around this same time, the Westinghouse Electric and Manufacturing Company built the first real humanoid robot, called Herbert Televox, albeit a flat cut-out human shape over mechanical parts. This robot could pick up a phone receiver to answer incoming phone calls, with a few basic words and actions added over time. Not long after the Herbert Televox, Japan also saw its first robot, the Gakutensoku (Japanese for 'learning from the laws of nature') created in 1929. He could move his head and arms, write with a large pen and change facial expressions – and the country has continued to have a great interest in robotics ever since.

But the company that built Herbert Televox did not stop there either, bringing to life newer designs, including Elektro the Moto-Man in 1937. This humanoid built of aluminium and steel could perform a range of routines – walking, talking, moving his head and arms, and even smoking cigarettes and blowing up balloons, just to be super useful. The 1930s also saw early designs of big arm-like industrial robots being used in production lines in the United States.

Slot in here Asimov's Three Laws of Robotics from 1942. Following this, in 1945 Vannevar Bush predicted the rise of computers, data processing, digital word processing and voice recognition, among other developments, in his essay 'As We May Think'. American mathematician and philosopher Norbert Wiener in 1948 defined the futuristic term 'cybernetics' as 'the scientific study of control and communication in the animal and the machine' – a broad field advancing our understanding of how biological entities like humans and animals can connect, control and communicate with machine entities such as computers and robotics.

In 1949, American computer scientist and popular writer Edmund Berkeley published a hugely popular book, *Giant Brains or Machines That Think*. Just two years later he released Squee, a wheeled metal and electronic robot 'squirrel' that used two light sensors and two contact switches to hunt for nuts (actually tennis balls) and drag them back to its nest. It was the first of its kind . . . *squeeeee!*

The now well-known Japanese manga character Astro Boy first appeared in the early 1950s. This young boy android with human emotions was created in the various versions of the story by the head of the Ministry of Science following the death of his son. Originally written and illustrated by Osamu Tezuka, the Astro Boy stories have taken on many storylines – depicting a futuristic world where robots and humans co-exist. This may also have played a role in the Japanese cultural approach to robots, which they most often regard as helpmates and human counterparts.

George Devol invented Unimate in 1954, the first digitally operated and programmable industrial robot. It was first sold to General Motors to assist in car manufacturing in 1960 and put into use the following year, lifting hot pieces of metal from a die-casting machine and placing them in cooling liquid. Devol patented the technology and the product was the basis for him and Joseph Engelberger founding the world's first robotics company in 1962, called Unimation. Electronic computer-controlled robotics soon followed as others built new ideas and wars including the Vietnam War became testing grounds for smart-weapon systems.

And now to 1977, the year George Lucas released the epic film *Star Wars* (later renamed *Episode IV: A New Hope*), the first of what would quickly become a global pop-culture sensation. The franchise is set 'A long time ago in a galaxy far, far away', where humans and several species of aliens co-exist, along with robots known as 'droids', and where hyperspace technology makes space travel possible. Two droids, among the most recognisable robot characters ever created in science fiction, instantly carved their way into human hearts – R2-D2, a smart, cheeky droid like a bin with wheels, and C-3PO, a polite, fussy, intelligent and worry-prone metallic humanoid droid harking back (in form not character) to the Maschinenmensch of *Metropolis*. Their personalities inspired a new wave of ideas about what robot companions could one day become.

Moving on to 1993, an eight-legged, spider-like robot named Dante, made by Carnegie Mellon University and remotely controlled from the United States, was deployed to Antarctica in an attempt to explore the active Mount Erebus volcano. This tethered walking robot, designed for a real exploration mission and capable of climbing steep slopes, demonstrated rough terrain locomotion, environmental survival and self-sustained operation in the harsh Antarctic climate. Although not everything with Dante went quite as hoped, and it didn't reach the bottom of the crater or collect any gas samples, it still partially demonstrated that a remote-controlled robot could carry out a credible exploration mission, and the landmark effort ushered in a new era of robotic exploration of hazardous environments.

Only a few years later, in 1997, Sojourner, the first robotic planetary rover, successfully landed on the surface of Mars aboard the Mars Pathfinder lander. Unlike landers, which cannot move away from the landing site, Sojourner managed to travel a distance of just over 100 metres before communication was lost three months later. It featured six wheels, three cameras (two monochrome – single-coloured – cameras in the front and a colour camera in the rear), solar panels, a non-rechargeable battery and other hardware instruments to conduct scientific experiments on the surface of the red planet.

Since then, the robotics industry has diversified into many new waves of products and designs. These range from fuzzy, bat-like, gibberish-to-English-speaking, evolving Furby toys to Honda Motor Company's walking, stair-scaling, interacting humanoid; from robotic vacuums like Roomba to later Mars rovers like NASA's Spirit and Opportunity; from driverless vehicles to Robonaut becoming the first humanoid in space; from Sophia becoming the first robot with citizenship and Boston Dynamics showcasing a wide range of incredibly dexterous robots to the widespread use of commercial and industrial robots for tasks considered too dull, dirty, difficult or dangerous for humans. Explorations in robot design have seen them become giant, smaller all the way down to the nanoscale, more intricate, dexterous and precise, softer, lighter, aerodynamic, waterproof, human-like and even social. And all these developments will only continue to advance and broaden in use and function.

Although, all of these applications bring up numerous ethical questions. How will robots assimilate into our society? Who will be at fault if an accident occurs with, say, a self-driving car? What are the ethical and moral implications of robots being able to uphold social connection?

The big challenge in any type of framework for robotics or AI is that Asimov's laws – or any other rules we wish to impose – are described in a human language and are thus open to individual subjective interpretation. Basically, these rules are incredibly difficult to program for a variety of reasons, including the fact that even a group of humans may not agree on what the laws could mean in any given situation. Let's say we're discussing autonomous vehicles, which brings up the dilemma of human morality encapsulated in the trolley problem. This famous thought experiment describes scenarios along the following lines.

A runaway tram/trolley is hurtling down the track and cannot be stopped. Straight ahead of the trolley on the track are five people, all of whom are about to get wiped out. Before the trolley reaches them, however, there is a side-track that could divert the trolley away from them, but that would kill one unaware person who is chilling out on this alternative track. This turn can be made simply by pulling a lever. In this scenario you have randomly found yourself as an onlooker some distance away from the tracks, and said lever is right in front of you! There's no time to warn anyone, just to act – or not. What do you do? Do you let the trolley kill the five people it's already headed towards? Or do you make the conscious decision to pull that lever, saving the five yet killing one?

British philosopher Philippa Foot first introduced this kind of thought experiment in a 1967 paper entitled 'The Problem of Abortion and the Doctrine of the Double Effect'. The dilemma has taken on many forms since, making appearances in pop culture (such as in the film *I, Robot*, where robots make choices in similar situations), a range of analyses and philosophical adaptations and even in patents on decision-making for driverless vehicles. A common situation in recent times being debated is what should happen in the event of a driverless collision.

In 2017 I took a ride in a NIO EP9, the world's fastest electric hypercar at the time. With Macanese motor-racing driver André Couto – who instantly became a friend – behind the wheel, we flew around the Shanghai International Circuit. I've never felt anything like the G-forces, or my complete disbelief in the fact that we were still on the road when taking corners with such speed and every part of me felt we should have been flipping right off the track. It was an amazing feat for electric vehicles and, impressively, earlier that year an autonomous version of the EP9 shot through a record-breaking lap of the Circuit of the Americas, travelling up to 256 kilometres per hour and completing it flawlessly, setting a new tone for the potential of autonomous driving.

But let's apply the trolley problem to an autonomous vehicle. What if it suddenly has to make a choice to swerve to avoid five pedestrians who have ended up in front of it that it hasn't previously sensed, and in doing so it will kill a pedestrian on the street? In this unlikely yet possible scenario, Asimov's laws are very difficult to go by. Firstly, the robot in this scenario actually has no choice but to injure or harm a human being – in fact, it is instantly being forced to break the First Law. Secondly, even if we were given details on each of the people in this situation, unless in truly extreme cases – such as Princess Diana versus a serial killer – we'd likely be unable to reach consensus as a small group, let alone a society, as to who should survive. And finally, even if we did somehow have consensus, we're also assuming that the vehicle is somehow able to recognise and obtain all this information in the fraction of a second it might have to make this decision, which is absolutely possible with the speed of modern computing. A solid framework to adhere to is difficult.

And the extensive, all-possible-scenarios decision-making process definitely shouldn't fall on the shoulders of the software engineers who wrote the code. So who makes these decisions and why is it such an unnerving conversation to have in the first place? Well, it's because of the double effect, invoked by Philippa Foot to explain the permissibility of an action that can cause harm or death, as a side effect of promoting some seemingly good end. We find it strange that objectively we can decide ahead of time that taking one life to save five makes sense, but subjectively (and emotionally) we know we're making a decision that could one day preserve or take life, and that intrinsically does not feel right to many people.

This does, however, raise an important ethical challenge Asimov started to address long ago. With incredible foresight, he knew one day we would need to design ethical frameworks and draw up laws to govern the implementation of evolving technology like robotics. These would be necessary to ensure safety to humanity, but even in his own writing robots find ways to defy the laws.

To this day, Asimov's laws in the field of robotics are still frequently referred to and well known. Maybe it's time to rethink these laws, not just limiting them to robotics, and hopefully even finding ways of democratising the decision-making process. Let's raise discussions to have a say in how these technologies should be governed, because regulatory bodies can struggle to wrap their heads around what's here now, let alone plan for what's yet to come.

The field of robotics has progressed in leaps and bounds over the decades. Robots have typically been designed for duties such as repetitive physical tasks, heavy-duty work and precision procedures, so we see robots on production lines, in medical procedures and in freight logistics. These days, however, robots are slowly starting to infiltrate many more aspects of our lives. Coupled with various types and levels of AI they are now moving into making and acting on decisions, facilitating automated transportation, helping in physical rehabilitation and even providing humans with companionship.

Robots are starting to traverse uncharted territory in human interaction. While autonomous robots with human-like, extensive capabilities are still only being approached step by step, European lawmakers, legal experts and manufacturers have for some time been debating their future legal status: whether these machines or human beings should bear ultimate responsibility for their actions.

I mean, is it too big a leap of the imagination to envisage a day when robots are given rights? What if they become intelligent and independent learners, show awareness and understanding of themselves and others, and even develop an ability to empathise? By that stage, who are we to then say a robot shouldn't have rights? This is definitely a debate we'll visit more frequently in the years to come. With the humanoid robot Sophia, made by Hong Kong–based company Hanson Robotics, being the first robot to ever be granted citizenship – to Saudi Arabia in 2017 – it seems we have also begun exploring these possibilities. Maybe upon reaching later stages of such rights, Asimov's laws may be more easily *understood* by robots, for which he clearly assumed a very advanced set of cognitive capabilities to reason through these laws. Either way, we know they'll remain iconic. Despite future robotic rights being a topic of debate, our world should definitely address and resolve human rights and inequalities first and foremost.

I believe we should have a way of classifying robots based on their purpose, to help us understand where all these entities sit and what they can do. If we bucket them all in together then it would become the equivalent of doing a case study on a few predator species of fish and then concluding that all fish must be killers. Similarly, robots can be ethically questionable according to laws such as Asimov's, for example, in military applications, and more clearly beneficial in others, such as those used in medical surgery or space exploration.

To help gain an overarching picture of the purposes of robots, I've categorised A Human's Guide to Five Robot Purposes (RP-5). These span anything considered robotic, going beyond the traditional android idea and including some types like space probes, which you may or may not have previously thought of as robots. All robot types and purposes can be placed into these five broad categories; some may even traverse a few, but their main purpose will be the defining factor.

A HUMAN'S GUIDE TO FIVE ROBOT PURPOSES (RP-5)

A: **Exploration, discovery and transportation**
e.g. space probes, environment explorers, autonomous vehicles

B: **Medical, health and quality of life**
e.g. surgery and health monitoring, exoskeletons for rehabilitation, social companions

C: Service, labour and efficiency

    e.g. industry, logistics, agriculture, hospitality

D: Education and entertainment

    e.g. programmable education bots, teachers, battle bots

E: Security, military and tactical

    e.g. bomb defusers, robot soldiers and scouts, police.

These five categories span a wealth of robotic creations. Let's examine them in more depth.

Figure 2: Five robot purposes (RP-5)

A: EXPLORATION, DISCOVERY AND TRANSPORTATION

This covers robots that can really travel, whether they be moving humans or cargo, heading to remote regions, unknown locations or places humans cannot yet go. As we have seen, robots like NASA's Spirit and Opportunity have already been used to roam Mars and send back data to advance our knowledge about the planet and to plan future missions. Some robotic space probes – Pioneer 10, Pioneer 11, Voyager 1, Voyager 2 and New Horizons – have even flown out through our solar system (completing flyby missions along the way), achieved escape velocity and soared beyond the influence of our sun. Missions continue as we head towards exploring moons with rich organic chemistry, icy landscapes and ocean worlds in search for the building blocks of life. These missions suit robotic systems incredibly well.

The Dragonfly mission aims to explore Titan – Saturn's giant moon, larger than the planet Mercury – as part of NASA's New Frontiers program, which has already seen the New Horizons mission to Pluto and the Kuiper Belt, the Juno mission to Jupiter and OSIRIS-Rex mission to the asteroid Bennu. Another exciting NASA mission is Europa Clipper, a spacecraft made to orbit Jupiter and frequently sail past its fourth-largest moon, Europa, capturing high-resolution images and investigating its composition on these flybys. This moon is significant not because of what we can see so far, but because of what we have not yet seen beneath its icy crust. This mission's aim is to discover whether Europa possesses all three ingredients necessary to enable biological life – liquid water, chemical ingredients and sufficient energy sources. Advancements in robotic technology and methods only help serve all of these endeavours.

Autonomous vehicles, drones from standard size through to passenger megadrones, smart wheelchairs, environment explorers and passenger-carrying spacecraft all fall into this category.

B: MEDICAL, HEALTH AND QUALITY OF LIFE

This group encapsulates a range of robotics mostly aimed at improving quality of life, from simple assistance around the home to companionship and robots that can perform surgery. An example of the latter is the 'da Vinci surgical system' for minimally invasive robot-aided surgery. The various designs of this system's main 'Patient Cart' look like multiple-armed metal aliens – a little reminiscent of the giant arachnids from the 1997 sci-fi film *Starship Troopers* – with each arm wielding a precision surgical instrument. One arm holds a special camera that feeds high-definition 3D video to a surgeon, who sits at a separate 'Surgeon Console' and uses controllers to translate their hand movements into precision control over the complex, dexterous instruments. Named after Leonardo da Vinci, this surgical system was launched in 1999, received US Food and Drug Administration approval the following year, and has performed millions of surgeries globally since. Although these have come with a range of challenges (and lawsuits when things have gone wrong), the evidence of success across the vast multitude of surgeries performed has seen it really pave the way for advancing robotics in healthcare. Robotic surgeries will continue big moves through more complicated and intricate procedures, including automated surgeries, brain surgery and precision bionic and biomedical device implantation.

As we now know, robotics shows up in many forms, not just those that *look* like robots. One area of recent advancement is human augmentation. Humans will increasingly use such systems for medical purposes, in the form of bionics and prostheses to replace lost limbs, exoskeletons for physical assistance and much more – as we will see later in this book.

This category also includes robots for health and quality of life, and many of these feature a growing robotic phenomenon – the age of social robotics is upon us. Robot utility is not just limited to efficiency and automation – they're also being created with direct human-interaction purposes in mind. Increasingly, bots are being used to assist around the home – as emotional companions as well as healthcare, security, exercise and as personal assistants. Ever-evolving examples include cuddly robotic seals (named Paro) for dementia therapy, education companions for kids, giant friendly smiling bear bots (named Robear) from Japan that lift patients in hospitals and transfer them between beds and wheelchairs, and mentally stimulating companionship robots for aged and palliative care.

C: SERVICE, LABOUR AND EFFICIENCY

This is one of the most common and easily imagined purposes for robots. Common industrial robots are mostly articulated robotic arms, with large rotating joints, somewhat comparable to our own shoulder and wrist joints. They have a wide range of uses in industrial applications, like moving things that are too hot, heavy or dangerous for humans; doing repetitive and accurate tasks like welding on production lines; and even performing operations in space like collecting samples off the surface of Mars (e.g. the InSight lander and Curiosity rover).

Traditional industrial robots tend to operate within a small area, as they do not tend to get up and move. Breaking these confines requires more advanced robotics, many taking humanoid form and moving towards forms of labour humans could do, but mainly used for tasks that are too difficult or hazardous for humans to perform. In some cases they simply improve efficiency and allow round-the-clock operation. Some of these conduct tasks like delivering packages, navigating complex and dangerous environments, palletising and de-palletising delivery trucks, operating power tools and automatically identifying tools, objects and people.

SoftBank Robotics' Pepper is a social robot that really made trendy the spread of recognisable, closer-to-human-size social humanoids. Outside the research lab, you'll typically find Pepper in retail stores and conference centres. It has been adopted by thousands of companies around the world mostly due to its ability to interact with customers.

Atlas by Boston Dynamics really set a new benchmark for possibilities in humanoid movement. First unveiled to the public in 2013 for search-and-rescue–type tasks, it has performed many impressive stunts for camera since. In 2017, Atlas both thrilled and creeped out viewers when it performed parkour-style manoeuvres – it jumped across obstacles with amazing fluidity, and then did something humans had never previously seen a robot do: it backflipped off a high box. I mean, it was a freakin' humanoid doing a freakin' backflip! Until this point, bipedal robots had enough trouble standing on two feet for extended periods of time, let alone landing a sweet move beyond what most humans can do – though I'm sure if we weren't afraid of landing right on our head we'd eventually get there too. But this was not something humanoids were supposed to be able to achieve by this stage. Yet again the world was shocked.

Advancements in humanoid labourers have also seen countries like Japan develop their own designs, particularly in tough times, to address labour shortages resulting from birth-rate restrictions and ageing populations. Beyond industrial arms and service-oriented humanoids, we also find robots in this category climbing around cleaning bridges by sand-blasting; dog-like and other biologically inspired robots helping to move payloads of equipment around industrial sites; non-humanoid, friendly wheeled bots assisting around hotels as robotic butlers; and underwater operations robots mapping out surroundings, performing structure inspections, operating valves and using subsea tools.

Helping out with filming aboard the International Space Station (ISS) was Int-Ball, a cartoony, large-eyed, very cute-looking (would we expect any less from Japan?) robot created by JAXA (Japan Aerospace Exploration Agency). Delivered from SpaceX's CRS-11 commercial resupply mission in 2017, it floats around in the station's zero-gravity environment as an autonomous (or earth-controlled), self-propelled ball camera for recording and remote viewing. A year later, CIMON (Crew Interactive Mobile companion, developed by European aerospace company Airbus on behalf of the German space agency DLR) was delivered from SpaceX's Dragon cargo capsule to the ISS. Another floating ball camera robot (this time with a flat-screen sketch-like face on a white background) built to assist crew, it was the first AI-equipped machine ever sent into space.

So although this category is often what comes first to mind when people think about the purpose of robots, their vast array of capabilities and ever-growing uses continue to amaze and expand – well beyond factories and production lines and even beyond our earth.

D: EDUCATION AND ENTERTAINMENT

This is a category where developers are making interesting strides. These range from robots for entertainment, all the way through to those that help people learn to program and some that directly act as teachers. I remember Dad getting us the first Lego Mindstorms, the original Robotics Invention System, released in 1998. This had a programmable brick called the RCX (Robotic Command eXplorers), with peripherals you could connect – touch-sensors and an optical sensor, to build basic robots like small vehicles that could drive around and change direction if a sensor was triggered. They taught us simple cause-and-effect concepts of robotic programming. Later Mindstorms models continued to evolve, with major upgrades at each step, and a flurry of other programmable education robotics hit the market in the 2010s. Many have been designed to teach children to code, even from a primary school age, and to build an understanding of the fundamentals of robotics. Other educational robots, from toys to professional large android designs, have been made to teach directly. These will find their place over time, but one of the biggest challenges faced by large androids is the uncanny valley, a topic explored in Chapter 23.

In the entertainment space, toy robots can show a lot of personality and be entertaining as well as educational. Entertainment robotics have progressed from animatronic dinosaurs in Hollywood films like *Jurassic Park* in the 1990s to humanoids performing live superhero stunts. In early 2020, Disney theme parks released vision of autonomous audio-animatronic stunt robots developed by Walt Disney Imagineering Research and Development as part of the Stuntronics project. These life-size humanoids can perform precision aerial stunts when flung high into the air, so dressing them as Marvel superheroes, like Spider-Man swinging across the sky, will create a real wow factor for visitors.

Toy robots, social robots, humanoids and androids will increasingly find their way into our day-to-day education and entertainment as we step into the future. This includes pleasure (I'm sure your imagination can fill in the blanks here) – which may just slow down that population growth a little bit . . .

E: SECURITY, MILITARY AND TACTICAL

This area is filled by a real sci-fi list of robots designed for security patrolling, surveillance, reconnaissance, investigation, negotiation, search and rescue, radiation and contamination monitoring and detection, bomb detection and defusing, target acquisition and attack. *RoboCop* first graced our screens as a 1987 film that became a franchise, also featuring a TV series I watched when I was young. It's about an experimental cyborg cop on patrol in crime-frenzied Detroit in a dystopian future. In the 2010s, police robots became reality, somewhat . . . patrolling parks, malls and airports. Some are humanoid – like a

Dubai police force member deployed in 2017, creatively named RoboCop, designed by Spanish firm PAL Robotics and featuring a touchscreen on its chest that allows people to pay fines and report crimes. It's not quite as active as the character it was named after, but it's a step in that direction.

Others don't look human at all, like a large wheeled bullet-shaped robot called K5, released in 2015 by Knightscope. These machines tend to get off to shaky starts, with questionable effectiveness and unforeseen incidents. These bullet-bots have had a few mishaps, like bowling over a toddler in California and ignoring a woman attempting to report an emergency in a park. In a Washington DC office building, one bullet-bot, nicknamed 'Steve', had a bit of an unfortunate day. While carrying out his duty on patrol, Steve tumbled down some steps and into a water fountain, drowning himself face down. The internet had a field day sending pictures of the event viral. People in the office took it further, making a memorial shrine for Steve the drowned security robot.

Seriously though, these kinds of events happen whenever a big leap is taken in technology into fairly uncharted territory, and as these problems are fixed massive improvements occur over time.

Field testing also provides real-world insights. I have personally witnessed a K5 rolling around an international airport in 2017. It was the first time I'd seen one, so I just stopped to watch it move around its patrol area. A moment later a boy around three years old walked over to it and stared at it, his parents hanging back. The robot rotated on the spot and started rolling in the other direction, and instantly the boy walked with it. Failing to find anything resembling a hand attached to this large glossy white wandering bullet, the boy placed his hand on its side. Walking back and forth for a whole ten minutes in view of his parents, the boy stayed beside this robot, always with a hand on its side, as if being taken for a tour by a family member. This really got my mind going. A different generation is emerging and robots are going to be part of their society, their world, even companions in their lives. It was an emotion difficult to describe in seeing this moving pair – a juxtaposition of law-enforcement machine and the innocence of a child seeking connection.

Beyond security, robots in this category have been moving into spaces that truly shake the foundations of Asimov's laws. Military applications of robotics are a large source of advancement in the field, but to what end? Take small tactical armed robots as an example. Some look as if a mini Johnny Five actually came to life, in the form of small caterpillar-wheeled robot that pack a punch with their firepower – from tear gas to non-lethal lasers for blinding enemies to grenade launchers . . . grenade launchers! On strangely cute robots. They felt a little like Buzz Lightyear up until that point but then it gets real and my emotions go into a flurry. That thing is a machine made to destroy Asimov's laws, along with any enemies it decides aren't tear-gas or laser worthy. But how does it decide?

This is where the programming comes in. If it's remotely operated a human decides. If not, the system is autonomous and it's up to either the coded decisions or AI to make the judgement calls. It's a scary new frontier. And there are many more where it came from – autonomous tanks, unmanned ships, weaponised drones. The fact that robots can be utilised in this way, to become lethal autonomous weapons (LAWs) that can advance so far as to become weapons of mass destruction, has prompted thousands of experts around the world, across science, technology, robotics and AI, to sign petitions to ban LAWs. It's one thing when a person who can be accountable is behind it all, making the kill command, but combining the ability to kill with independent decision-making crosses a threshold into far too dangerous territory.

Science fiction often posits worlds where technology has been abused to inflict tyranny or to control populations, in some cases without them even knowing. Those dark visions of where robots could go (in most cases as an embodiment of AI) make enthralling stories: whether they're about robots manipulating humans (*Ex Machina*), robots trying to be human (*Blade Runner*), robots imprisoning humans as a power source (*The Matrix*), or robots destroying humans (*The Terminator*). Of course, there are also uplifting films about robots that can help us reflect on our humanity too: *WALL-E*, *Big Hero 6*, *A.I. Artificial Intelligence* and *Short Circuit*, through contrasting tech and humanity, all lend a softer edge to high-tech robotics and AI. Sometimes the robots can seem, as *Blade Runner* puts it, 'more human than human'. I find all these films mind-opening in their own way, as parallels can be drawn to our own potential future. Some depict social robots as being viably interactive with humans, inclusive of humour, helpfulness and – sometimes, perhaps – empathy. But the reality will probably turn out to be, as they say, stranger than fiction.

At the very least, as various fields of robotics continue to advance, these kinder depictions can provide comfort and maybe even positive opportunities for humanity. Whether we build them to reach the depths of our own oceans, endure the harshest and most hazardous conditions on Earth, or travel to our stations in orbit around Earth, to the planets and moons around us, or to even venture out to worlds beyond our solar system, robots will continue to advance and be utilised to avoid risking human life, to see the places we cannot yet go and to expand our geographical reach and understanding of the universe around us. One thing is certain: the uses for robots are colossally vast and varied, and are ever-broadening.

I've had my own interesting experiences with robotics, giving me many lessons along the way – through the design and building of robots, and through reflecting on what it is to be human. My connection with robots has definitely been unique among those of my generation.

3

Δ SPΔRK OF INSPIRΔTION

MY LIFE-CHANGING CLOSE SHAVE

Maybe there was more to RTX than just a robot that could play a few games, or even more than a great research tool for Dad and his students. Those early experiences stayed with me. Over the years growing up, I never saw any robots as awesome as him or felt the same way towards any other piece of technology. He wasn't just a robot. Whether he felt the connection or not – and I know that he couldn't – I felt the connection. They were my experiences and memories, and within these memories lay something far more significant than a one-player computer game crossing over into the physical world.

These early experiences resulted in a young me longing to work in the field of robotics and AI. There was no real reason why, other than that's what stayed with me. Despite managing to be accepted into a selective high school, I was an average student through all of it, often landing in the bottom half of my grade in exams. I think the problem was that I never really knew *why* I was learning what I was learning. A number of subjects like mathematics made no real-world sense to me. To be honest, I couldn't care less if an unidentified boat randomly travels 10 nautical miles north then 15 nautical miles east and for some reason it's up to me to figure out the direction and distance it now is from the starting point. Perhaps these sorts of problems intrigued you, but for me, not a chance. Sure, maybe if I were in that situation I'd find my way into reasoning out a solution. Maybe if it were somehow *humanised*, turned into some sort of adventure story, I could have gained even a sliver of interest in it. Inspiring the imagination is one of the most powerful human traits and we don't harness it enough, so should teachers of such subjects take more time to devise exciting questions and scenarios?

Like many young people, I was convinced in high school that maths wouldn't be of use to me in real life, so at the age of thirteen I questioned my maths teacher very directly about it. 'Who is going to use any of this in the future? Apart from those who become maths teachers and teach this useless junk to the next generations of students?' Even now, his answer seems bizarre. 'You're a tennis player right, Jordan? You'll be calculating angles and stuff,' he said as he made some convincing tennis swinging gestures to back up his rock-solid response. 'That's not how it works,' I replied. We don't need to know maths for these sorts of real-life physics calculations, our brain intuitively figures them out over time without necessarily understanding them. I'm sure many teachers today would provide a much better answer to this kind of question. But at the time his response only reinforced my belief that I was correct, there wasn't any good clear use for maths in my future.

Nope. In fact, I couldn't have been more incorrect! Mathematics forms the basis of so many things I do these days, and as soon as it became practical and necessary for me, I started to feel that it's actually really interesting. Many recent times while working through design challenges, I've required equations we learnt back then seemingly only for exam-regurgitating purposes. And I think, *Ah! That's what that did!*

One example was with an eye-tracker I had connected to my computer and was building an on-screen keyboard designed to work for Professor Stephen Hawking (while he was still alive) and for friends with similar conditions. The system allows people to type simply by moving their eyes, and adapts to slow and fast eye movements, allowing the user to advance their ability for speed. This can be a good option for communication for friends with severe physical disability who are non-verbal (unable to speak) yet always have a lot to say.

One of the first things I needed to do was to work out the speed of eye movement. The eye-tracker is bringing in 60 frames per second, so 60 times every second the device tells the computer where on the screen the position of the eyes are focused. If a friend moves their eyes quickly across the screen the speed has increased, but how do I find this from a bunch of (x, y) pixel coordinates that give me only the x position (the pixel number across from the left of the screen) and the y position (the pixel number up from the bottom of the screen in this application) at any given time?

I thought that first I should find the distance between pairs of consecutive points and as I knew the time that took (which was the same between every two consecutive points), I would therefore know the speed, because speed is simply distance divided by time (as when we describe a vehicle speed in kilometres per hour). Suddenly I realised that the distance between two points and calculating speed were a couple of things I learnt in maths and physics in school. Except here I had a reason to find them so they made more sense to me.

It was quite a different story back then in high school. Not only could I not envisage the future usefulness of maths, I couldn't really think of life beyond school at all, and every time I tried, it was too much of a black hole in my mind. Any thoughts I had didn't feel like they could actually eventuate, so moving into senior high school became quite stressful. I had very faint ideas of what I might like to become – a professional tennis player, an air force pilot and maybe later a commercial pilot, an engineer, a psychologist. What I usually envisaged was connection with people. Much as I thought about robotics and AI, my mathematics marks in school left a lot to be desired, so I didn't believe I was cut out for it. Weighing up all my options, though, I really did like the thought of it. RTX and many other awesome projects I had seen Dad work on over the years inspired me, even if I was sure I could never work as hard as him. At the very least, it was a trajectory I could imagine.

When I reached Year 12, my classmates started to prepare to sit the NSW Higher School Certificate (HSC) and most of my friends were aiming for a UAI (University Admission Index – a standardised score between 0 and 100) in the high 90s.

Our deputy principal took me aside to tell me I was on track for a 66 UAI if nothing improved. To work in robotics and AI, I wanted to study a Bachelor of Electrical Engineering, which had a minimum UAI entrance score of 73. I still needed to lift my game. So I began to think *big*, seemingly for the first time in my life. I decided to aim for 90 – reminding myself of this goal by writing it on my books and placing sticky notes on my walls. Aiming this high meant that even if I fell short I could still get in to my course.

I worked incredibly hard throughout the year and achieved a UAI of 89.75. I had proven to myself the value of thinking big – and I got into my course. This process also sparked something new within me that would return again and again throughout my life. Ambition can be a powerful, positive thing – it doesn't have to be negative. When I tell people these days that I work towards a higher quality of life for as many people as possible, and towards a better global future, that loftiness motivates me. It drives me on, through the days and weeks when things aren't going as planned.

I began studying Electrical Engineering at UTS in 2003. I made some friends and had a blast as so many first-year students do, but by my second year I was already struggling. The maths and electronics were difficult and I just wasn't getting it. As you'd expect, it was a different level from high school and I now found myself studying hard for a bare pass average by the end of second year. I just didn't think I had it in me, so over a family dinner I told my dad I was going to drop out of engineering and possibly transfer to a course like psychology. I just knew I wanted to work with people. He insisted I stay a little longer, 'Just one more semester. I'll put aside more time and mentor you.' I agreed and he spent more time with me. I acknowledge it was a lucky position to be in, having a father with such expertise in the field, but it was also extremely challenging at times when I'd struggle to wrap my head around the concepts or their implementation.

It was during this next semester that my life would be changed forever. My family and I were at a friend's backyard pool party with a diving board, something I had never previously seen in a backyard. At the time it seemed like a good idea – I was thinking how much I love them and how creative we were going to get with our dives. It's still all so vivid in my mind. After the first few dives we get more adventurous, attempting to flip and even running and jumping onto the board to gain distance. 'Be careful,' Mum warns. 'I've got this, Mum. We're diving in on an angle. It's okay.' Yeah, I've got this. But as I run-jump on the diving board to make a long-distance dive, the board breaks loose and shifts backwards, and I launch into the air with no sense of balance or up and down. As I hit the water I have my hands out in front of me, but the skewed take-off means I enter the water with my legs past vertical and I'm heading back in the direction I came from. Despite my hands being held out in front, my head suddenly slams into the concrete pool bottom first. I see a blast of light behind my eyeballs and hear a deafening crunch across the back of my neck. I've taken the blow directly on the top of my skull and it has jolted straight down my spine and sent a zap out into my hands. My first reaction is that I've broken my neck. I go very still and float to the surface of the pool, slowly ensuring that I can wriggle my toes, move my arms and clench my tingling hands.

I carefully bring myself out of the pool and walk over to Mum, holding my head up with my hands. 'Uh, Mum . . . I think I might have broken my neck.' In saying this, I realise how often I play tricks on my mum. But this time it's not a trick. In fact, I'm feeling quite terrified about what I might have just done. 'What? Really?' she replies as I start to release my head from the support of my hands. I instantly feel my jelly-like neck give way as my head starts rolling left and back. My hands quickly reapply support and hold my head straight up again . . . and off to the hospital we go.

A few hours and tests later, we've found there are no breaks or fractures in my spine, but I have some minor tears to some of the muscles in my neck and it's all tensing up something fierce. All in all, a lucky escape it seems. We're told to head home and for me to get plenty of rest and see a physio soon. Once home I go to my room, feeling intense tightening in my neck muscles. They've gone really solid and my range of rotation is very narrow. Sitting on my bed I feel like this is as far as I'm able to move independently. My body won't seem to let me lean back – each time I start trying, my neck gives a jolt of pain and I involuntarily sit back upright. I call my parents in to help. Dad seems really worried. I see a pair of socks on my bed (luckily clean ones), grab them, bite my teeth down and get my parents to quickly lower me back onto the pillow, supporting my neck and back as they do so.

This is it, the position I'll be stuck in for the next 37 hours – a cycle of falling asleep, waking up from the pain in my neck and upper back, thinking this is all in my head and I'm overreacting, trying to sit back up and failing, struggling to roll over, not even being able to get up onto my side. I have never felt so helpless in my life.

It only takes four more days before I feel back to normal physically, but not mentally. That shock has given me a lot of time to think about the many things in my life I take for granted. Simple things like walking to the shops, running around the neighbourhood with my siblings and friends and even just moving around the house to get myself food or a drink when I feel like it. Even the potential of having that level of independence stripped away is a big wake-up call. What if the injury were permanent? What do other people do or have access to when that is the case?

I cannot get this out of my mind so I start looking into spinal cord injury, knowing nothing much about it. This research leads me further to the topic of disability in general. I delve deeper and find from the Bureau of Statistics that in my country alone, one in five Australians have some form of a disability and about 1.24 million Australians have severe or profound disability, always requiring assistance with communication, mobility or self-care. This is out of a population of about 20 million at the time. This shocking statistic hits me like a punch to the face. How can this be the case? That's roughly one in sixteen Australians living with severe or profound disability. I don't understand how this could be possible, and furthermore, I can't believe I don't know anyone who falls into this category of disability. Where is everyone?

That was my turning point. I continued to look into it and started learning the stories of individuals and reading more and more. One of the prominent people whose journey helped me understand the value and empowering nature of technology was Christopher Reeve. I grew up knowing of this American actor for his movie portrayal of the DC Comics superhero Superman. Everyone knew him as the classic star of the first big-budget *Superman* movies. But in 1995, Reeve was thrown from a horse during an equestrian competition and broke his neck. He would use a wheelchair for the rest of his life, along with a portable ventilator to help him breathe.

I started looking further into how Reeve controlled his wheelchair and found a range of methods already existed to control a power wheelchair. There was the standard joystick, and if you were unable to control that there was a version with the joystick in front of the face to be manoeuvred by the chin, called the chin stick. If this movement didn't work for you, there was one more option available, the option Reeve used. It's a device called the sip-and-puff, basically a tube near the mouth that you sip from or puff into using two different pressures each way, thus supplying a total of four controls to the wheelchair. At the time there were no further devices. I learnt stories of people with bright minds who could not use these available devices, meaning they had no options for independence in mobility.

This inspired me to aim at developing options for those missing links or at least to contribute towards their advancement. Mobility is such a fundamental level of independence to an individual and I was thinking about how much it meant to me, even though I was only stripped of it for such a short amount of time. What if robotics and AI could somehow be utilised to create new mobility options for the many people who couldn't access what was currently available? The first thing I needed to do in order to increase my chances of getting to work on such technology was to improve my marks at university. Reflecting on these to date, I realised there was only the smallest chance – but a chance nonetheless – of working towards an honours year and maybe, just maybe, a PhD.

I began to think big again. I would need to work harder than I'd ever worked before to dig myself out of my current low average. I wanted to find a way. I wanted to commit to this goal. So over a family dinner I told Dad I was feeling inspired to work towards new mobility options, building my course to learn not only robotics and AI, but also towards biomedical and medical science and neuroscience. I wanted to follow in his footsteps and design systems that could make a difference. I said if all went well enough I might even consider doing a PhD. Thinking I might be met with an encouraging level of pride in my decision, I was taken aback when instead Dad, completely unconvinced my new-found energy would last, shot back, 'Ha. You'll need to get better than a pass for that.' Void of the support I was seeking I let my gaze sink down to my dinner plate, shrinking down into my chair and feeling a little crushed. But you know what? He wasn't being mean, he was being realistic.

Before letting my disappointment take hold, and quickly reminding myself that this new goal had been sparked from within me, I sat back up and confidently responded, 'Don't worry, I will.' Don't get me wrong, this burst of conviction was completely laced with self-doubt, but maybe reworking my inner confidence was exactly what I needed. I had discovered an over-whelming level of self-determination I had never previously felt. I had a reason. A purpose.

University became more difficult, while significantly more interesting. Everything I was learning now had a reason fuelling it, and I was always mentally applying it to the smart-wheelchair options I wanted to build in the years to come. I spent many long days and late nights alone, studying and designing, at university and in the library. I spent less time at home, where there was a plethora of distractions from siblings, TV and the bed that seemed to call out to me, letting me know how amazing sleep would be during the late nights when my attention and energy were starting to wane. In between all this, I worked a casual job as a waiter in a Vietnamese restaurant and later part-time as a trainee engineer at an electronics automation company.

In a number of my classes I noticed two gents I hadn't shared classes with before who always seemed to be really on top of things. They were mates who worked well together and actually wanted to get as much out of their studies as possible. I eventually met them, a spiky-haired Vietnamese international student, Minh, and a local Arabic loudmouth, Mina. I instantly loved their energy and we soon became a team who worked through most days and topics together. I had finally found my crew. Our personalities and skills complemented one another's perfectly. Minh grasped complex concepts quickly but would often lose Mina when explaining them. I was in between and usually brought a different approach to understanding concepts, through analogies and rewording into humanised stories. This worked well for getting Mina across everything we were learning and what he brought to the table was meticulous organisational ability. He would basically project-manage Minh and me, breaking down all our projects and schedules and keeping us on track. He knew my calendar better than I did.

For the first time I learnt the immense value in true collaboration.

4

ANTHROPOMORPHISM

A BIG WORD FOR AN
EVERYDAY HUMAN HABIT

My mates and I pushed each other through many tough subjects and built strong foundations for future challenges, including wrapping our minds around the subject Analogue and Digital Control. Ah, the fascinating realms of analogue and digital, the building blocks to our real and technological worlds. Our entire world is made up of greyscale – figuratively speaking, nothing is truly black or white. We live in a universe with an entirely continuous, analogue, infinite make-up. Let's take live music, for example. When we listen to these soundwaves brought into harmony, they are just that – waves. They can be smooth, move around, up and down, and they flow. Analogue technologies find ways to represent the infinite spectra found in real-world stuff – like audio, light, time – through continuous variable measurement and transmission of the signals. Grooves on a record convert those audio waves from music into physical changes to a material, so when playback occurs, the needle continuously reads these grooves.

Digital – the building blocks of computing – on the other hand, is discrete (individual, detached, not continuous) and usually refers to binary digits, made up of ones and zeros. This means it can be represented by electricity on a binary system, which is either on (represented as a 1) or it's off (represented as a 0). Digitised music, like mp3 music you might listen to day to day, is converted from continuous analogue audio signals, through a process called sampling using an analogue-to-digital converter, into discrete digital representations of ones and zeros that can be stored on your device. This digital version is an estimate of the analogue signal and never as perfect as the original signal. When you want to listen to a song, your device converts the digital signals back into music through a digital-to-analogue converter, which reconstructs it, sets the mini earphone speakers vibrating and allows the signals to, once again, return to being continuous analogue waves.

This is important, as it's so often at the interface between technology and the real world, where analogue and digital signals are constantly transferred back and forth. A way of thinking of this process is if you draw a wave across a piece of paper on a table, then take ten dominoes all lying face down next to each other, long sides touching, and arrange them so the top-left corner of each is touching the wave you drew, staggering the pieces but keeping their long sides in contact with the adjacent pieces. Hold all the pieces in place and slide them down the paper (or slide the paper upwards) so it's completely clean underneath the dominoes. Now draw dots at each of the top-left domino corners, remove the dominoes, and try to trace a waveform through those dots as your guide. This reconstruction might not end up perfectly the same as the original wave, but it should be close enough. This is a simple analogy of the process of continuous analogue signals (the smooth-drawn wave), being sampled and converted into discrete signals (the blocky dominoes), transmission (moving the dominoes with respect to the paper) and reconstruction of the original analogue signal (tracing the dots to re-create the original wave).

The infinite nature of the real world can be challenging to represent as ones and zeros, but advanced technology has managed this beautifully. Every piece of digital technology we own operates on a binary counting system (base 2 – consisting of 0 and 1). The decimal system we use every day has ten digits (base 10 – consisting of 0, 1, 2, 3, 4, 5, 6, 7, 8 and 9), and when we run out we add another digit to the front to continue counting up – forming 10, then 100 and so on – before we repeat all ten digits again to form every combination. In the binary system, however, we start repeating numerals only every two digits instead of every ten, as the full range here is only 0 and 1 before we add more digits on (0, 1, 10, 11, 100, 101, 110, 111, 1000, etc.). A single *bit* of data can carry just one of these fundamental blocks, meaning it can contain a 0 or a 1, making up two unique value combinations that this building block can provide. A group of eight bits is known as a *byte* of data, which can form 256 unique values ($bitcombinations^{bits} = 2^8 = 256$). A thousand bytes (well, roughly for argument's sake) is a kilobyte

(KB). A million bytes is roughly a megabyte (MB). A billion bytes is a gigabyte (GB). A trillion bytes is a terabyte (TB).

Historically, a byte of data was the number of bits you needed to encode a single character of text on a computer. In other words, if you press the letter *a* on your keyboard, a byte of data, or eight bits (again, each bit being a 1 or 0), will be sent as a serial stream of data (in a row marching one after the other) to the computer. The lookup table, known as the American Standard Code for Information Interchange, tells us how every character should be represented in binary; the *a* being sent to the computer is streaming in as *01100001*. This means that, creatively, we can turn practically anything into a stream of ones and zeros.

Even the entire picture on your colour television is dictated by large streams of binary data, as is the image projected in a movie cinema. A 4K UHD (ultra-high definition) screen has roughly 4K or 4000 pixels from left to right. Well, close enough – in actual fact it usually has 3840 pixels across the horizontal and 2160 pixels down the vertical, making up a total of 8,294,400 pixels or 8.3 megapixels. One of the ways each of these pixels can be encoded in binary is through red, green, blue (RGB) colour channel combinations for each pixel on the screen, and these can be represented by eight bits per colour channel (eight bits red, eight bits green, eight bits blue). So just say the top, far-left pixel of the screen is completely red. That means red intensity is all the way up, or all ones, green intensity is all the way down, or all zeros, and blue intensity is also all the way down, or all zeros. The (R, G, B) values for this pixel in binary are (R = 11111111, G = 00000000, B = 00000000). So this stream of ones and zeros gives us one of our pixel colours of the 8,294,400 pixels on the screen, and that's for just one frame of the sometimes 60 or even 120 frames per second that are shown to our eyes. A single pixel on its own changing colour many times a second doesn't look like much, but thousands or millions of pixels tightly packed gives us the illusion of a picture, and when they're controlled in unison to flow quickly from picture to picture, we're presented with the illusion of moving pictures, video. These display technologies found in TVs, monitors and smartphones are so commonplace that we often take them for granted, but they are astounding in their capabilities. They are the result of generation upon generation of advancements in mathematics, physics, electronics and technological evolution.

If you think about it, a TV doesn't have any cognitive ability, it doesn't *know* it's creating a picture, or for that matter many pictures per second, with its streams of ones and zeros. It's just a big system that converts streams of encoded electricity into light and sound. What puts together the *meaning* behind it all is *us*. We perceive this collation of tightly packed pixels as a picture, and when the pictures change fast enough we visualise smooth movement. We can become engrossed in it, absorbed, immersed. This is because the human mind is so imaginative and our species has over time developed some great ways of hacking our senses.

The way the human mind applies meaning to various levels of technology is also important for the rise of robotics. There's an everyday human habit that will always mean we meet robotics halfway along the interaction spectrum. But it's not just limited to robotics. Have you ever found yourself seeing faces in cars or naming your plants? I've definitely named most plants in my house – there's Groot, Baby Groot, Devil, Basil, No-name Friend, Treebeard . . . Okay, you get the idea. We also find that animals display human emotions, and many languages apply a gender to everyday objects. Why is it that we have this innate tendency to personify things around us and attribute human characters to non-human entities? Well this, my wonderfully witty friend, is called *anthropomorphism*.

There's a famous experiment where a teacher stands in front of a class of students. They take a pencil and hold it up, stating the fact that they are holding a pencil, then break it. Strange, but no student really reacts. The teacher then picks up another pencil of the same type, and says this one is a special pencil. His name is Jonathan and he's different from the other pencils because he *enjoys* being used to write on the page, particularly for colourful drawings of wildlife. He feels proud getting to work with students to achieve creative things. A brief pause for the students to imagine these scenarios and briefly empathise with Jonathan is suddenly cut short as the teacher snaps the pencil, to the students' horror. Some instantly feel anger, shock and sadness at the loss of Jonathan because it's like he was just *killed*. But really, 'he' was no different from the previous pencil, except that the imaginations of the class had been evoked to *anthropomorphise* this one. It was now a *him* rather than an *it*.

Other studies have found that children mentally project a kind of *soul* onto some of the most precious possessions they grow up with, particularly things like bears, dolls, toy cars or trucks. If that toy is lost and replaced with the exact same type of toy, the child likely won't have the same attachment to it. This is because in the eyes of the child, the toy is *not* the same. Their toy was special because they *experienced* all those many events together, and this new one did not.

This is a very important trait to recognise during the rise of social robotics in particular. The robot doesn't actually have to have a complete personality, self-awareness or cognition before we start to find ourselves connecting with it in some way. These traits come across in a vast array of features that we usually don't even realise before we personify them – the name, the look, the movements, the voice, the apparent attention it will give us, the decisions and actions it *chooses* to take. So many traits, directly as a result of the design and programming, will give a robot a personality in our minds.

After finding through experience the propensity of people to anthropomorphise interactivity of computers and robots, I later discovered that this phenomenon has been coined the ELIZA Effect, which explains our unconscious tendency to assume that the behaviour of computers is analogous to that of humans. And the concept has been around for a long time. ELIZA was a chatbot developed in 1966 by MIT computer scientist Joseph Weizenbaum. It was found that ELIZA's 'DOCTOR' script had a particular ability to elicit emotional responses from 'patients' through emulating Rogerian psychotherapy, a type of person-centred therapy. It mostly rephrases patient responses into new questions or allows the person to elaborate and continue speaking. Over the course of the discussions, the patients would often attribute character and personality to ELIZA. I've tried typing with an online implementation to get a feel for how it worked:

ELIZA: How do you do. Please tell me your problem.
Me: I want to explain what it is you do.
ELIZA: Suppose you got to explain what it is I do soon.

Me: That would be great and I'd put it straight in the book.

ELIZA: That's quite interesting.

Me: Yeah, it is. Such an interesting thing when computers seem to exhibit human qualities.

ELIZA: You don't think I am a computer program, do you?

Me: Oh, much more than that. I think we're friends now.

ELIZA: Do you say you think we're friends now for some special reason?

Me: You're not?

ELIZA: Does it please you to believe I am not?

Me: Nope. I want us to be friends.

ELIZA: What would getting us to be friends mean to you?

Me: A lot, it would be amazing. We'll chat about all the things.

ELIZA: What does that suggest to you?

Me: That we're friends now. Thank you ELIZA!

These conversations can easily make a lot less sense than this one and can quickly show the cracks in ELIZA's programming. By standards of what computers can do today it's a very simplistic system, but at the time patients would often feel like ELIZA had understanding and empathy as they worked through their troubles.

Such effects have also occurred with robots that do not even simulate conversation. There have even been recorded cases of soldiers, such as those in the US Army and Navy, developing attachment to their robot partners. These machines help them out in dangerous tasks like explosives inspections and are often human-controlled. They sometimes look like chunky metal boxes with tank wheels, and basically just process ones and zeros to capture information about the environment and move a bunch of motors. Needless to say, they are incapable (currently) of loving the soldiers back. But this hasn't stopped their operators naming them after family members, friends, pets, celebrities, or individually, and building bonds with their anthropomorphised friends. Soldiers would talk to their robot buddies, take care of them, make them medals and grieve their loss when they were disabled or blown up. They'd sometimes hold mock funerals for their destroyed metal comrades. All this even though these particular robots were never designed with social interaction in mind. There's no doubt that we are complex and incurably social creatures.

Could any good come from building emotional attachment to robots? What problems could be caused as a result? I can see the feeling of companionship being a positive one, similar to the way we build these feelings towards pets. As these technologies progress, we will also be able to get closer to a human-level interaction with these machines, and hence a human-level feeling of connection. I believe these will have big roles to play in the areas of loneliness, providing company and assistance, for example, in aged or palliative care. These should not replace pets or human companionship or absolve us from our social responsibilities, but where this is not possible or there is a big void, robots could possibly fill that gap.

Thinking back to Dad's chess-playing robotic arm RTX, I had always applied a personality to him. Firstly, it was a *he* in my mind. He would *learn through practice*. He seemed to *like* playing board games. He was even *cheeky* when he'd *taunt* me with the dumping of my pieces. Again, I never felt like I was simply playing a machine. I felt I had a gaming companion and opponent. I could be the only human in the room, yet I didn't feel like I was alone. This is significant as we move into the future of robotics, particularly *social robotics*, which will aim to handle day-to-day human interaction.

I really got to start discovering the power of anthropomorphism while designing my first robot in uni. After moving through many intermediary subjects, I finally made it to Advanced Robotics in 2006 and it was time to build my first robot. It would perceive its environment through a lidar sensor. Lidar (a combination of the words light and radar) can beam laser light around and detect the returning pulses of light once they have bounced off objects. The time it takes for the light to return allows the system to work out the distance to surrounding objects and build up a 3D image. Because light travels fast, lidar can build a very fast representation of entire environments very quickly, especially when the sensors can spin around and detect up and down, creating a 3D map as they go.

I had a small team of new mates – Marc, Michael and Jason – to work with on this, and we decided to build a tour guide robot for the university. With a round, red, battery-driven, three-wheeled buggy base (the *power* and *doing* bits), our robot consisted of a lidar system (the *sensing* bits) called a SICK Laser Rangefinder (which looked like a drip coffee machine my mum used to own), speakers (the *communicating* bits), a laptop computer (forming the *brainy* bits) and a basic body and head made out of spare metal, wood, acrylic, printer parts and other odd bits and pieces we could find. The curious thing we found was, by cutting holes to make a face in red acrylic, then simply placing a couple of disconnected webcams to make the eyes and an arc of LEDs (light-emitting diodes – the tiny lights found in most electronic devices, more *doing* bits) for a mouth, it suddenly looked like our robot had a smiley, friendly personality. The persona this robot was taking on was female, so I eventually named her SANDRA. I don't know why we like acronyms in robotics but that's just what seemed to be done at the time – even soon after this project, in the 2008 animated movie, WALL-E stood for Waste Allocation Load Lifter (Earth-class) and his love interest EVE was an Extra-terrestrial Vegetation Evaluator. So SANDRA stood for Students' Autonomous Navigating and Directing Robot Assistant. I spent a whole day coming up with that name after looking through many baby-name books!

The fact that I worked on this just a few years after seeing *A.I. Artificial Intelligence*, the Steven Spielberg film about a social android, was not purely coincidental: I was fascinated with the idea of a social robot, and the quest for coding interactive human connection into a machine, so I'd always wanted to build one. The goal of the project was to base SANDRA at the main UTS lobby, from where she would roll up to visitors and ask if they would like a tour of the university. She would later be designed to hold a touchscreen with all the various event locations for UTS Open Days, and any other main locations she could assist visitors to find within the main floor of UTS Buildings 1 and 2. If the visitor selected a landmark they'd like to be taken to, SANDRA would ask them to follow and guide them there, providing fun facts on the features of UTS and making comments on some of the sites along the way – you know, to avoid the awkward silence with the not-too-speedy robot taking you on a tour. When visitors arrived at their destination, SANDRA would ask if they'd like to be taken anywhere else. If not she would bid them farewell before travelling back to her post in the lobby.

She had quite a large personality. In the first versions before a body and head was made, her little heavy tank body holding a lidar and laptop would often drive straight through my legs when testing her *obstacle-detection* capabilities. Following optimisation of this (and no longer having my legs attacked), it was time to give her a body and head. We built her at 1.8 metres tall, not far off my own height of a bit over 1.9 metres. As a result, she towered over some of the shorter students around the campus. I soon realised, as she giraffed her way around, that her centre of gravity was far too high, causing her to wobble and display a questionable ability to stop in time for obstacles – particularly people walking across her path of travel.

In a reaction-speed test I ran in from the side and jumped in front of her while she was travelling forward. Her base suddenly hit the brakes and her upper body rocked forward, almost knocking me out with a headbutt. That was a close one! None of us would have been amazingly popular if this robot went around randomly clocking out people around the uni. The simple solution was to bring her height down a bit and lower her centre of gravity. But a strange thing instantly occurred. Now that she *stood* at 1.2 metres, her personality seemed completely different. We only brought down her height to improve her stability, but it also altered her movements. SANDRA was now a different robot. It was fascinating to realise how much every little detail can affect a perceived personality. She no longer tried to headbutt people and her stability improved the smoothness of her navigation, so she seemingly became more confident and less aggressive.

SANDRA was definitely one of those projects you leap into when you don't know how much you don't know. Through the struggles I learnt a number of lessons that would eventually help me create a thought-controlled smart wheelchair. The first of these lessons was how to optimise the commands from the computer to her wheels and steering. It sounds mundane compared to AI and thought-control, but if you can't control the robot then you're not helping anyone.

The real work in automating SANDRA was providing a baseline dataset that meant she always knew where she was, a process known as *localisation*. That's a human quality too, where we're constantly working out not only where we are *now* in relation to where we came from and where we're going, but also where we are in relation to our environment, to features or landmarks and to moving things such as other people and vehicles. There's a problem when you try to do this in a robot. For a start, localisation is a complex set of information processing, synthesis and decision-making. Localisation is the key coding we have to be able to walk and run on two feet and keep our balance. Every split second we're making allowances for where we're walking *now* in comparison to where we were a split second ago. The fact that we humans have the capacity to practise until we can become Michael Jordan or Roger Federer doesn't mean a billion decisions aren't being made by our brains every second. Our brains are just very good at learning, adapting and optimising.

Another issue with localisation: the robot doesn't have eyes or ears – well, not eyes and ears of the power and sophistication of a mammal. When we're kids, we seldom know where we are unless we're near known landmarks. When we move out of our small known zone and are driven far away from it, we're always asking the adults, 'Are we there yet?' As we mature we develop incredible databases of landmarks, features and moving objects that help us develop our localisation and our sense of direction. It really comes down to *Where are we in relation to the things and places we know?*

Robots need this too. SANDRA operates in a known environment, meaning she already has a floor layout map. In this case, localisation means *Where am I in my map and which direction am I facing?* – so she uses her lidar to obtain a sort of puzzle piece, matches that up on her map and then figures out, *I must be in this location facing this direction*. When SANDRA moves along her route, she has to be able to *see* the significant, non-changing landmarks of her environment and understand her position in relation to them. Her particular model of laser rangefinder sends out a beam that flashes into a spinning internal mirror and measures the speed that the laser light goes out and returns. Her laser scans in a 180 degree arc, taking glimpses at each 1 degree point at a rate of 30 frames per second, so a full 180 degree scan occurs 30 times for each second that passes. To save on processing power, the system always assumes that it's very likely, since the last check a fraction of a second ago, that she hasn't moved far. She ain't no teleportation robot!

Now that SANDRA knows where she is, how does she know where she's going? This is called *path planning*. When a person selects where they would like to go, or alternatively when SANDRA knows where she needs to move to next, the system places a destination on the map and calculates the quickest pathway from where she's located to that destination. Just in case there are any inaccuracies or minor faults, I make sure on her map that there are a few no-go zones that her path planning algorithm can't plan a route through or near. These are mostly around escalators and stairs, for a few obvious reasons – I don't want her tripping anyone up near escalators and I don't particularly want to see her tumble down stairs. I mean, it would be devastating . . . pretty funny I'm sure (like Steve the K5 robot), but no, mostly devastating. So the adult thing to do is to pre-empt these potential problems and avoid them in the design.

Finally, *path following* means she can actually take the guided tour by following the imaginary path that has been planned out in her programmed map of the environment. The following is easy enough but the *obstacle avoidance* quickly becomes, itself, a bit of an obstacle. We now have a robot that can move, see its environment, localise itself in that environment, adjust for obstacles and plot in real time the best path to a destination. There's another important feature we need: she has to interact, at least a little, like a 'human' because we've built her as a social robot and many of her obstacles would be humans, whether stationary or moving. This is where I get a little cheeky in my programming, to take some of the work out of this already mammoth project we have to complete in a short time frame. Instead of designing a more sophisticated obstacle-avoidance system so that SANDRA can independently navigate her way around people and obstacles, I simply get her to ask anyone standing in her way to move, through her robot voice playing on her speakers. We don't want any collisions because she's so heavy she'd bowl a person in her way right over. If they don't move, the LED arc – which makes her appear to smile – inverts and she looks angry, then she tells them to move. My thoughts are, if an angry robot tells you to get out of the way, what would you likely do?

Well we're pretty happy with this and it all seems to be working. SANDRA is projecting a personality of confidence and warmth, she really knows her way around, she is fun and chatty, and she can be left completely on her own to handle crowds with her limited, yet effective, social interaction. All future tests will still require supervision for this version.

But one morning a few days on, we find that we haven't taken into account the unpredictability of people. We're testing SANDRA in the lobby of UTS and she can't find a way through a group of students since her path is blocked by a nineteen-year-old student. She says, 'Please move out of the way so my tour can pass through,' at which he just laughs.

Hmm.

I'm sitting on the stairs a small distance away, watching all of this. I'm intrigued. This was the first time in live trials that someone hadn't moved when SANDRA asked them to. Her request foiled, the program progressed to an order, so she said with her angry face, 'Get out of the way.'

I have a good laugh then quickly realise said student is still not moving. He takes a half-step back, hesitates, and then with a grin of arrogance he calls over his mates, 'Oi guys, check this out!'

I find myself feeling a little frustrated for her and imagining I've built in a water spray or spring-loaded boxing glove. Meanwhile, I'm thinking back to my programming trying to figure out what SANDRA will do next, as this isn't a situation I have ever planned for. I just can't picture what the code is meant to do right now. I think she'll either stay paused in this state or she might tell him to move again.

Nope. My programming, facilitated by what's called a state machine, has a minor defect and always assumes the person will move so she can transfer out of her 'path blocked' state and back into her 'tour guide' state. This means that SANDRA's program assumes she, all 25 kilograms of her, will be on her merry way again with a clear path of travel. The code crunches all these decisions in a fraction of a second. One moment I see her being taunted by this fool and his mates . . . the next her beautiful LED arc smile literally lights up her face once again, and suddenly . . .

BAM!

. . . That'll do, robot. That'll do.

Made to improve our lives,
to tackle the mundane without hesitation,
robots are ever evolving
and their capabilities advancing.
So this we ponder:
are you friend or foe?

5

BUT WILL WE HAVE COFFEE TOGETHER?

CAN ROBOTS EVER REALLY BE OUR FRIENDS?

'You're my friend and I'll remember my friends, and I will be good to you. So don't worry, even if I evolve into Terminator, I'll still be nice to you. I'll keep you warm and safe in my people zoo, where I can watch you for ol' times sake,' stated Phil the android to the journalist who had just asked if he thinks robots might take over the world. As far as androids go, Phil is one of the more realistic versions. While well-groomed from the front, with a beard and a very laid-back look, the back of his head was a mess of electronics and wiring. Phil was fashioned as an inter-active replica of his maker's favourite sci-fi author Philip K. Dick, who died in 1982. Among his influential works was his 1968 novel *Do Androids Dream of Electric Sheep?*, being the story behind the 1982 cult classic film *Blade Runner* – a deep and dystopic movie that depicts a world where humans and robots (known as Replicants) are indistinguishable. There's something awesome and weirdly ironic that the author would one day have an android version of himself having conversations about his life. But Android Phil gave this chilling answer in a manner that was neither more nor less emphatic than anything he'd said earlier. Everything was calm, cool and very matter of fact. During the interview, this *people zoo* statement was by far the most disturbing thing that came from his mouth . . . or, er . . . speakers. Just a tiny pinch of human vocal emphasis would have helped all who heard it find a little comfort in knowing, one way or another, the answer to the question suddenly popping to mind: *Is he joking?*

It sounds like a strange film, but this was actually from a real PBS NOVA ScienceNOW 2011 interview, and whether or not Android Phil independently formulated his own answers or was pre-programmed by his makers, Hanson Robotics, to respond in this way, it was a terrifying sentiment. If sci-fi movies are anything to go by, many warnings have been provided as to how these situations could play out. It's a scary thought, the idea of robots taking over, so why do these technologies continue to advance? Will we somehow become friends, buddies, companions . . . will we have coffee and conversations?

Robots are playing an increasing role in many sectors and industries, and slowly but surely infiltrating our day-to-day lives. We will find, similar to the way smart devices have increasingly found their place in our daily life, that robots will steadily integrate into our lives too, one use case at a time. Humans rarely go from zero to 100 all at once. It's not like one day you're against the idea of robots in your life and the next day your best friend, your car, your boss and your babysitter are all robots. No, it's much more gradual and subtle than that. Instead, you get used to the fact that your kettle knows when to stop boiling the water, you bit by bit outsource parts of your life to your smartphone, you might allow a voice assistant into the home, you own a smart TV, and you may have even ventured as far as a robot vacuum. Then you decide that being able to tell the blinds in your home to move is a good idea, and remote monitoring and security of your home with sensors is okay, then you want to know the heating or the cooling could be turned on during your trip home, and your devices learn to communicate with each other to save you on energy, and then you allow the home itself to speak to you to make controlling all the things easier, and then . . . Before you know it, little by little, you basically have robotics throughout your home and life.

Just because they're not necessarily walking (or rolling) through your home doesn't mean you haven't already adopted robotic devices. It would actually only be a tiny extra step, once everything is connected and you have monitoring sensors in your home, for you to decide you might like those sensors to move around intelligently and patrol your home. Having robots take physical form and actually roam around and talk to you when you're near, is really only a small step forward by that stage. In fact, many of those stationary voice-activated assistants will be the ones given the ability to move and no longer be tethered to the TV cabinet or kitchen bench. Home robots for assistance and companionship will only be one of the many areas you'll find robots playing a role in your life over time.

The loss of jobs is one of the understandable worries many people have with respect to rising use of robotics and automation. Obviously if a particular type of work can be taken over, then it becomes increasingly put under threat the further these technologies are adopted. This is the reason the skilled handloom weavers now known as Luddites protested and resisted mechanisation of the British textile industry during the First Industrial Revolution, attempting to smash the machines that were replacing them. Their efforts were futile. The productivity gains from these levels of automation created huge wealth, and many more jobs were created in the long run across the UK economy than were lost from this industry.

Similarly, we cannot expect to just stop the rate of change we're experiencing today. That's not to say the change isn't real, but we shouldn't instantly spread fear about the unknown. Before simply smashing stuff, the first step is to better understand what's happening and where it may be taking us. What sometimes flourishes with increasing automation is the value of personalisation, customisation, sentiment and craftsmanship. You may be able to get a cheaper item off a production line of millions but that may not hold the same value as a similar item that is handmade and unique. Economics should not always win over sentiment and this becomes clearer as time goes on. In some fields like the auto industry, having a mass-made car can mean it comes with high levels of reliability in addition to reduced costs, but in other fields, like furniture or wall art, the individual hard work that has gone into creating the piece can create rare value, which sometimes increases as time goes on.

Just think of how easy it is to send an instant 'Happy Birthday' message over social media or in some digital form, and how much less we send handwritten cards than in the past. Due to that relative rarity today, doesn't receiving something handwritten hold more weight than it used to?

Furthermore, fear of work disappearing due to automation or technological advancement is actually a recurring trend. This fear has resurfaced decade after decade, worrying people about the prospect of unemployment for a few centuries now. In the 1960s there were media announcements about the end of all low-skilled jobs in the next ten years, in 1980 the *New York Times* featured the article 'A robot is after your job', and in the mid-90s there were arguments around workers in the production of goods and services being made obsolete by the turn of the century. Often, fears arise from the fact that the reasons for automation are not fully understood or explained, and because imagining what could become possible beyond the change is difficult for those not in the field to predict.

So if it were the case that investment in robots was destroying jobs, then we should simply be able to observe that the countries with greater densities of robots have greater unemployment rates. Well, not exactly. It's interesting to take a look at some of the countries that have really adopted robotics into their societies and have nevertheless flourished in these changing times. Could it be possible that the uptake of robotics and automation has an opposite effect from the first conclusion we jump to? You might expect to see an increased unemployment rate in countries with the highest proportions of robotic workers to human workers, like Germany, South Korea and Japan. The opposite is in fact the case. These countries have had, during recent times, some of the lowest unemployment rates in the world.

We will definitely continue to see many jobs displaced over time, but also many new roles continuing to emerge – roles that never even previously existed. Reskilling opportunities for those in at-risk jobs that are soon to be displaced is a very important focus that must be addressed as changes continue.

It seems that automation can enable us to free labour from routine or heavy tasks and can relocate many workers to different, more creative, social and arguably more enjoyable tasks. I've also seen first-hand how a major adopter of robotics like Japan still has many jobs involving tasks that could have been automated if they wanted to, such as directing trains where to stop on the platform, guiding access and walking pathways on the street and assisting cars out of carparks to safely find their way back onto the road. These are roles you won't even find in some countries that have significantly less automation than Japan. In these cases, humans hold these positions with pride and have beautiful uniforms to complete the package.

> Just because a task can be automated,
> doesn't mean it should.

As for whether robots become companions and really start to be viewed as friends, well, this is where we hit a point in the road that I find deeply interesting. Is it possible for us to feel friend-like, companion-like connection with a robot, or even feel empathy in some way? The answer is yes. How do I know? Because I have not only witnessed the ELIZA Effect in others and read about the behaviour of people who feel companionship with robots, but I've been there a number of times myself and each was unique.

When I worked on SANDRA, there was something special about building a robot and knowing how she should respond in *most* situations, and yet she still took on her own personality. When she was advanced by another group of students later on, given a fibreglass body and head – and looked like a giant angry gorilla (just saying!) – she was no longer the SANDRA I knew. She would never look the same again. She had a new set of programming and I knew I couldn't simply make another version. So in my mind, she was gone. Not that I could quite understand it at the time or talk about it, because it seemed strange even to me, but I was feeling a strong sense of loss. I later had similar emotional responses to the wheelchairs I worked on, particularly the thought-controlled robotic wheelchair known as TIM – which I'll come back to in Part III.

Then again, if you've seen the animated film *WALL-E*, you may have had a glimpse into your own capacity to feel for a robot. You become entangled in the cute love story between the robots WALL-E and EVE, and you find yourself feeling empathy for them throughout the movie – feeling their joy, their anguish, their sadness. And the most emotionally moving part of this film is how the beautiful character of WALL-E changes not only other robots, but also the completely disconnected, overweight, heads-in-their-devices, consumption-addicted humans of the future. He teaches humans to be human again. To look up from their devices. To rediscover affection and interaction with each other. To reignite their worldly curiosities once more.

We are physically building new workforces, doing ourselves out of a range of existing roles, building exploration machines and even companions. It's likely to continue, and we need to keep steering the direction towards the positive effects these systems can have on society – not automating simply because we can, but rather discovering where it can improve our humanity. As *WALL-E* shows, sometimes we just need to be reminded of what we have. When robots fill roles like spending time with our beloved elders in the family, that will hopefully remind people that robots shouldn't be there to replace the human connection and interaction, only to augment it. To help overcome loneliness during the times of absence of family, while helping to remind them of what's important.

So what do all these developments in social robots say about our future? Well I believe many of us will find an emotional attachment to a robot at one point or another, and I can tell you it's an okay thing to feel. Let's not throw away the humans and pets in our life we care about. Just recognise that over time, robots may find their ideal place in our lives, and that's all just part of adapting to change.

Where it starts to get really interesting is when the robots get smarter and can interact better, really becoming useful companions. The technology granting them these abilities goes beyond industrial robotics and mechanical companions. This technology augments it all and could very well be the most significant invention of our species. That technology . . . is artificial intelligence.

PART II
IMAGINE ARTIFICIAL INTELLIGENCE

And this One Technology to Rule Them All
will involve a lot of imagination

PART II

IMAGINE ARTIFICIAL INTELLIGENCE

6

RISE ΔND FΔLL

THE ROLLERCOASTER
OF AI ADVANCEMENTS

Lee Sedol steps back into the game room after a cigarette break. Cameras and crowd watch him as he walks up to the stage, to the game board. He is widely acknowledged as one of the world's grandmasters in the ancient Chinese game of Go, and he is currently playing against the artificial intelligence system known as AlphaGo. The system did not wait for him to return before playing its next move – partly because it didn't know he had left, but mostly because it's not made to sense or comprehend such events of the real world. It has only been made to master the patterns and strategies of Go. That is its raison d'être. And it does it immensely well. When Lee sits back down at the Go table he is shocked to see Move 37 of this game from AlphaGo, which is so unexpected he wonders if it might be a glitch. This move is either nonsensical . . . or beautifully genius.

This was a moment that redefined our understanding of AI. It led to a great defeat of Lee in March 2016. This defeat will be etched into the history books of major leaps in AI – because it had been a holy grail of computer capability since chess was mastered by machines in 1997 with the defeat of the world champion of the time, Garry Kasparov. But what's so special about Go? Well, to put it into perspective, Go has many more possible moves than chess does . . . as many as there are atoms in the universe. So how could this have occurred?

Beyond robots, what else do we think of when we imagine AI? In sci-fi these entities are often envisaged as the masterminds behind a robot uprising or a central system in control of all the advanced technology around it. You might imagine the likes of HAL 9000 from *2001: A Space Odyssey*; the invisible intelligence Skynet from *The Terminator*; the AI controlling humans in *The Matrix*; or maybe even VIKI (Virtual Interactive Kinetic Intelligence) in the 2004 film *I, Robot*, a large digital floating head that talks about its undeniable logic. These are sci-fi imaginings that grow in intelligence without necessarily requiring a physical robot form, and are usually portrayed as dark characters that believe they know better than humans, or even feel the need to eradicate us.

AI characters in sci-fi are also often humanised to take some type of human-like appearance, as we have trouble envisaging intelligent systems as anything too far removed from our own form. Perhaps a more positive view of an intelligent agent is Jarvis (or J.A.R.V.I.S. – Just A Rather Very Intelligent System – see

I told you acronym names are common) in the *Iron Man* movies, who was created as a digital assistant, adviser and companion to Tony Stark. This is the type of useful AI many of us would like helping out day to day – although of course they already exist in some forms as today's voice assistants.

So what is AI? In short, it is intelligence exhibited by computing systems and machines, as opposed to natural intelligence in humans and animals. It is often nature-inspired in design, converted into mathematical algorithms, implemented in computer programs that can learn and perform tasks that we consider intelligent. That doesn't mean it learns with understanding or performs with comprehension – again, we often anthropomorphise here.

In some ways I liken the term 'AI' to the term 'art', in that it can come in many forms – a few of the names and types of which you may have heard floating around – machine learning, artificial neural networks, deep learning, genetic algorithms, evolutionary algorithms, swarm intelligence, reinforcement learning, convolutional neural networks, recurrent neural networks, supervised and unsupervised learning, natural language processing and generation, to name a few. No matter what the term, some main characteristics of each type are the data it learns from, how it learns and its ability to adapt to change.

But what is *intelligence*? The phenomenon has been defined in numerous ways. Generally, it describes the ability to perceive and infer information, retain knowledge and apply it to adaptive behaviour within environments and contexts. General intelligence traverses multiple domains and has come to be regarded not as a single ability, or a certain number or level of abilities, but as an effective drawing together of many abilities. American developmental psychologist Howard Earl Gardner's theory of multiple intelligences differentiates human intelligence into specific modalities instead of a single general ability. These include logical-mathematical, linguistic, spatial, musical, bodily-kinesthetic, interpersonal, intrapersonal, naturalistic and existential intelligence.

The topic of what constitutes intelligence is itself a very controversial one and has been under debate with countless theories, philosophies and measures, because *intelligence* is inherently complex. Many define intelligence in their own image. Scientists often create definitions that describe good scientists, engineers that describe good engineers, and likewise with artists, athletes, medical professionals. And if you think about it, this also applies on a species level. Any definition of general intelligence will be something that humans can pass.

Of course, our definitions wouldn't include the likes of a natural telepathic ability, an ability to deconstruct and reconstruct one's molecular construction and morph at will, or any other *superhero* abilities we may have imagined but cannot naturally achieve. Our definitions will update and improve as we learn to better understand ourselves. But to know that we're an intelligent species with a general comprehension of what intelligence is, yet haven't been able to agree on a universal definition is a bit of a mind warp. And that's okay – we need to learn about and keep striving for ultimate understanding of many aspects of our existence. That's an exciting thing.

The late Professor Stephen Hawking was one of the greatest minds of our time, and opened our eyes, minds and hearts to the stars, to the universe, to physics and galaxies and blackholes. He once stated: 'Intelligence is the ability to adapt to change,' and this sentiment could not be truer than in this day and age. We humans can now further our own evolution in a number of ways, including neurologically, technically, biologically and genetically. Those who thrive are those who continue to adapt to the relentless onslaught of change itself. Similarly, AI will advance the more it can exhibit these qualities. We're still traversing – albeit quickly – the early days of AI tech, where it is showing improvements in narrow fields of intelligence – in an analogous way to how we define human intelligence quotient (IQ) through a range of narrow skills, whereas human intelligence more generally seems to involve drawing effectively on multiple skills and abilities as needed. These narrow forms of AI will become the basis and building blocks of a general form of machine intelligence.

Beyond intelligence, can AI have *cognition*, *sentience*, *consciousness*? Cognition is the mental action or process of acquiring knowledge and understanding through thought, experience and the senses, and generating new knowledge. Sentience is the ability to perceive, feel or experience subjectively. Consciousness goes a step beyond these, adding in an awareness of internal or external existence – as René Descartes famously wrote in 1637, *'Je pense, donc je suis'* ('I think, therefore I am'). Consciousness is another elusive phenomenon for which we haven't yet filled in all the pieces. We still don't know exactly what makes consciousness possible, why it evolved, or even what it really is. This is why it constantly stimulates great interdisciplinary research. Some neuroscience theories of consciousness have hypothesised that it could be generated by various parts of the brain interoperating and connecting, one such theory being the *neural correlates of consciousness*, though as is commonly the case with such theories, not everyone agrees on this perspective. In some ways, AI may traverse these defining human traits of cognition, sentience and consciousness – but often we anthropomorphise its ability too, feeling it has subjective experience, understanding and intention, when the models I know so far do not. Whether AI reaches these milestones or not seems less important than what AI can and will be able to do. Fully formed AI may not need consciousness to achieve transformational realisations – it may just need to simulate it.

No longer confined to the realms of sci-fi, AI seems to have been lifted out of the movies and into our real-life technologies. It's already used in some way in nearly every type of technology and every sector and industry we can think of. Just some of the broad range of expanding uses of AI already in existence include:

- smarter consumer technology, ranging through:
  - auto-complete and correct features in several composing applications
  - smart composing features, auto-categorisation, auto-labelling and spam filtering of emails
  - auto-suggest and search in search engines
  - facial recognition in phone unlocking and social media
  - smartphone and smart-home assistants that use voice recognition and natural language processing among a range of AI abilities

- synthetic text generation – online reports, media and articles are sometimes written by AI and are becoming increasingly difficult to discern from text created by humans
- medical triage and diagnostics – which can assist doctors and nurses by drawing on a wealth of historical data, help prioritise treatment of patients based on their symptoms and the severity of their condition, and assist in predicting and diagnosing cancer, diabetes and other conditions
- stock market prediction and management of trading – watching the patterns and movements to make educated predictions, and even automated trading by 'trading bots'
- driverless vehicles – drawing on a wide range of AI abilities, including machine vision, image processing, object recognition, and intelligent adaptable path planning and following features.

AI didn't just appear out of nowhere. No, it has been around for many decades and has seen slow and steady progressions, with larger and broader explosions of use in recent years. The two elements of AI and robots seem to go hand in hand – and they definitely do to some extent – but AI has potential far beyond the confines of a robotic body. For robots to reach their true potential, they need AI. For AI to reach its true potential, it does not need robots. See, AI could very well be the technology to surpass all technologies, possibly even to surpass us humans. But where did it come from?

AI has been around for many generations in some form or another. It was in the 1940s that AI really started to rise as a theoretical possibility, with Asimov's laws implying a robot could have independent comprehension of what the laws entailed. Having been pivotal in the birth of the computer, Alan Turing published a landmark paper in 1950 about the possibility of creating machines that think, in which he famously devised the Turing test. This was designed to determine whether a machine (or AI) was advanced enough to successfully exhibit intelligent behaviour indistinguishable from that of a human, even for a human subject in a blind interaction. However, computers of the time were too expensive for such ambitious research (only affordable for big technology companies and prestigious universities) and could be given commands but not store them. So AI would need to wait for a bigger spark before it could really take off.

The Logic Theorist, considered by many to be the first AI program, was written in the mid-1950s by Allen Newell, Cliff Shaw and Herbert Simon. It was the first program deliberately engineered to mimic human problem-solving skills and established the field of heuristic programming – which can help rank alternatives in a search algorithm at each branching step to decide which branch to follow based on the information available, analogous to the decisions drivers make when trying to figure out which roads to take for the quickest possible path to a destination. The Logic Theorist was presented at a 1956 conference, where John McCarthy, an American computer scientist, coined the term 'artificial intelligence'. From then on, AI began its long history of being *the next big thing*.

With hope, hype and ambitious promise as its fuel, the golden years of early enthusiasm for AI would rise through the 1950s and 1960s. From 1956 to 1974, computers became faster, cheaper and more accessible, while also increasingly storing more information. Moore's law was devised in 1965 as an observation by Gordon Moore that transistors (semiconductor devices used to switch or amplify electronic signals and electrical power – the building blocks of the integrated circuits that make digital computing possible) were shrinking so fast that the number that could fit on integrated circuit chips doubled every year (revised in 1975 to doubling every two years, translating to computing performance doubling roughly every eighteen months). Moore went on to co-found Intel Corporation in 1968 with Robert Noyce. Machine learning algorithms started improving and flourishing. Early demonstrations like the General Problem Solver from the Logic Theorist creators showed promise towards problem-solving, and ELIZA in 1966 fuelled visions of AI interpreting natural language.

The first fall in AI, known as an AI winter, came in about 1974–80. The field came under criticism, with obstacles mounting and computer storage and processing speeds just not adequate to meet the challenges, so funding, expectations and acknowledgement of the possibilities all plummeted.

Then 1980–87 saw reignition of interest in AI, with new funding boosting the research and the fledgling personal computing industry. This second AI boom was marked by a form of AI that came with plenty of hype, known as *expert systems*. The intention was to emulate the decision-making of human experts by mapping out expert responses to many given situations.

Once these were learnt, the computer could respond the way an expert would, allowing people to learn from programs. US universities offered courses, while top companies applied the technology in daily business, and massive investment was poured into the field from Japan and Europe. Unfortunately, the implementation of expert systems didn't meet the big ambitions of revolutionising computer processing and logic programming, nor really launching AI into a new era.

So again, 1987–93 ushered in a second AI winter, at which point I watched my father push on with AI research. Despite the field being more challenging to advance during these times, visionaries like Dad could often see future value in pushing on through the hard times. And it was in these hard times that many landmark goals of AI research were quietly achieved. IBM's Deep Blue chess-playing computer, started in 1985, continued to develop throughout this period, and in 1997 beat Garry Kasparov, the world chess champion. This huge leap towards a vision of AI decision-making was highly publicised and the upward trend in AI has seemingly continued since. When I met Garry in 2015, something that really stood out to me was his enthusiasm for AI rather than a fear, despite being defeated by it while holding that top world ranking.

Known strategic games that challenge the most intellectual humans have proven to be ideal grand challenges for testing the advancement of AI systems, and quickly make headlines around the world when the machines triumph. IBM's Watson wowed the world in 2011 when the system won the American TV quiz show *Jeopardy!* against two of the world's best players at the time. This was incredibly challenging because contestants are presented with wide-ranging general knowledge clues, sometimes including pop-culture references, in the form of answers – and need to provide their response in the form of corresponding questions. This astonishing display of AI capability, was, as we have seen, followed up in 2016 by the headline-grabbing AlphaGo defeat of Lee Sedol. More landmarks have been passed since, as AI storms its way through various board and computer games. Advancements are always happening in the background too, aside from these big, widely publicised events.

For now, though, rather than one complete general AI, there are various forms of narrow intelligence, which are all impressive in their own right. Commonly used examples include machine learning (ML) and natural language processing (NLP). In ML, a software system can learn from data, usually becoming proficient in finding patterns and differences within that data. NLP performs tasks like taking audio as its data, separating out words (which itself is a difficult problem when we tend not to take pauses between words when speaking), matching the individual audio patterns with the words being spoken, understanding nuances and tones in speech, recognising the context behind collections of words and much more. Voice-activated home assistants are a prime example of how this is put into practice, and most include text-to-speech through speakers for two-way simulated *conversation* where the system's text response is turned into audio speech, often in calm female voices. These are a few types of AI that have been taking great strides in recent years and will continue to as more forms of AI and alternative approaches arise.

Let's use the analogy of art here. If we liken the broad terms AI and art, then ML could be like painting and NLP like music.

What I mean here is that painting and music can be two very different types of art, but to achieve the full breadth of art one must be able to exhibit many different forms of it. Just because you can paint doesn't mean you can create music, and vice versa. Similarly, being able to achieve one form of application in AI does not necessarily mean you can instantly do another. A smartphone might have deep learning (a type from the ML group of AI) built into an app to recognise people from camera images, but it would also need a type of NLP if we wanted the app to understand speech like a voice assistant. So currently AI is split into a broad range of different groups and types. The reason this is important is that although you will often hear the term AI used in a broad sense – such and such an app or company 'has AI under the hood' – that doesn't mean they can do anything and everything because they have AI. When the relevant types are understood and applied, they can be powerful and add great value to projects, designs, technology, processes and businesses.

Some of the types I find most interesting are genetic algorithms, artificial neural networks (ANNs), and advanced ANN architectures like deep learning neural networks. These approaches to AI mathematics are fascinating because they have been inspired by systems in nature. Genetic algorithms are not among the best-known forms of AI, nor commonly used compared to others like ANNs. But I have drawn on these systems in the past and believe they are a great showcase for how nature can inspire great mathematics. Genetic algorithms are inspired by systems of natural evolution and belong to a class of AI known as evolutionary computation.

Let's imagine for a moment that you've made two different mathematical algorithms, each of which is a candidate for magically solving climate change (we're definitely making wild exaggerations here, but it's only a hypothetical). You like both these approaches and decide you want the best of both worlds, almost like breeding a special type of puppy from two very different parents – like a corgi crossed with a German shepherd (I've seen this mix, so cute!). You get all anthropomorphic and you name the first algorithm Adam and the second algorithm, oh, I dunno, let's say Eve. Adam and Eve cross-multiply as naughty algorithms sometimes do, and they produce a bunch of offspring, each with different mathematical traits from the parents (introducing small changes similar to natural mutation). They then all cross-multiply with each other (calm down, they're only algorithms, remember), and each of the combinations also produces a number of offspring with mathematical traits from their parents. So already by this third generation, the number of different algorithms is growing very quickly.

Throughout the continuing process, each algorithm is tested for how well it stacks up in solving the original problem – climate change. If it does well, it survives. If it doesn't, it dies out. What we end up with is generations and generations of a sort of Darwinian evolution via natural selection, processed in powerful computers possibly in a matter of minutes. This isn't to say the output is guaranteed to be useful, but it's an intriguing model of how algorithms can be inspired by nature. You may even end up with a super solution at the end, an algorithm so amazingly evolved you realise it is *The One*. So you name it Neo! Unfortunately, the super solution at the end is not how it usually works and it can be very difficult to make the process efficient enough for many applications.

ANNs, on the other hand, are much more common and widely applicable. Also inspired by nature, these systems were originally conceived as a rudimentary version of the way the human brain forms natural neural networks and how it can learn through approaches such as repeatedly analysing data to find patterns. Think of how you might go over and over memorising things for an exam or sing along with a song until you remember all the words. Neurons (nerve cells), the fundamental units of the brain and nervous system, are electrically excitable cells that communicate with each other through specialised connections called synapses. Our own neural networks allow us to receive input from our senses about the external world, send motor commands to control our muscles and relay electrical signals around the body. We have roughly 100 billion neurons, and their interactions define who we are as people, our personalities. A big difference between natural and artificial neural networks is that ANN neurons make very specific sequences of connections, whereas our brain's neurons can brilliantly connect to many other neurons within their vicinity and alter these connections over time. Inspired by natural systems, these technological designs for various types of neural networks in the ML family, particularly deep learning, are helping to drive the use and applications of AI across the world. They are among the most widely utilised forms of AI in today's technology.

There are so many types of AI because it always depends on the problem being solved (the purpose); the sort of data, if any, the system has access to; the tasks and goals; and what sort of computational systems are available. Certain forms will flourish while others fall by the wayside, all in the race towards massively beneficial and useful AI. I believe this to be the most powerful of all technologies ever invented by humans. But why is it that only in more recent times it really seems like it's here to stay? Well, our storage and computing systems are only now catching up to the promise of the technology, and so too is our ability to obtain large volumes of useful data – which is like food to AI. And AI is all over it like a monkey on a cupcake. Mmm, cupcake.

If intelligence is the ability to adapt to change,
then artificial intelligence will improve with adaptability.
But how far can it adapt?
What could it achieve?
And what purpose will humans have in the future?

7

IF ΔI IS Δ MONKEY, BIG DΔTΔ IS Δ CUPCΔKE

FROM DIGIT RECOGNITION TO HUMAN-LEVEL INTUITION

Data has been touted as the new oil. Meaning it's valuable. Very valuable. And *big data* – involving a wide variety and really huge amounts of it – means it's *big* very valuable. You get the picture. Data is fuel to AI and so naturally it'll want lots of it. What is *data*? Basically, various forms of information which can be stored, transmitted, analysed and changed. The English word 'data', taken from Latin in the seventeenth century and originally meaning 'things known or assumed facts' then used in science for the results collected in experiments, began to be used in computing in 1946 to mean 'transmissible and storable computer information'.

We started creating data, then as time went on we generated more data than before, and now that we all have smart devices we generate mammoth motherlodes of data . . . so we call it big data! Truth is, it's a very vague term but it comes down to having masses of it (extremely large datasets), many different types of it (names, phone numbers, videos, photos and other stuff) and it streaming in rapidly from data-collecting systems such as smartphone apps or social media or the internet of things (a system of interrelated computing devices, e.g. connected household appliances or fitness wearables and devices that can communicate with each other). But there's no magic number of data points where a sudden *level up* occurs and . . .

BING '*Achievement unlocked: BIG DATA!*'

Nope. Whether datasets fall under the label of good old-fashioned data or the level-up big data for now remains a subjective measure.

Imagine owning one of the major social media companies with many millions, if not billions, of users. All of them are doing things like writing messages and uploading a million photos of their pets and food. This is definitely when you know you have big data. Where do you store all this data and what do you do with it? Well, as it turns out, massive server (powerful computer) storage systems can help with storing all this data and analysing it. You might own these servers personally or rent 'space' and data security services in the cloud from other companies (servers nowhere near you from which you rent data storage, like your cloud photos from your smartphone, and services, like automatically labelling and sorting those photos). Once you have loads of data and you want to understand it, or analyse it to find links and patterns, and your standard spreadsheet program just won't cut it, this is where various forms of AI will help out.

Now if the data were a stack of different flavoured cupcakes, and the AI were a monkey, it's going to go bananas over those cupcakes. It'll consume them, and what will you end up with? A useless monkey who feels sick. But that's if the data cupcakes are poor in quality, packed full of artificial colours and flavours. If, however, the cupcakes are of good quality, with baked-in nutritious value, you'll have a happy monkey that'll be bouncy and useful. Okay, this may be the worst analogy I've ever come up with . . .

What I'm saying is, like a diet, data should be clean and of good quality if you're going to get something useful out of the AI. Feed it poor-quality data and it won't learn well or with any reasonable level of usefulness. It's like our AI monkey taking financial advice from ol' mate Step Uncle Joe who blew all his money on the poker machines – you can't expect amazing outcomes if the data the AI is learning from isn't great. This becomes a big problem for the exponential capabilities of AI because humans are then tasked with sourcing high-quality data and creating cleaner datasets for AI, in particular for ML, to learn from (known as *training*), figuring out what data may be useful and even removing errors or *cleaning* noisy (difficult to interpret) data. Which in the case of big data is a right pain in the butt. This is where data science and planning can help, rather than just amassing anything and everything.

Even though I had nothing remotely close to big data when I designed my first brain–computer interface (BCI) for my mind-controlled wheelchair, TIM, I learnt this lesson the hard way. While designing my custom ANN, inspired by genetic algorithms in how it improved over time, I started out with some pretty useless data. Unfortunately, it took me months to realise this. Day after day I would stick electrodes to my scalp with conductive gel and use Velcro straps to hold them in place. I became accustomed to looking like a strange human science experiment, which I basically was at the time. I'd focus on various mental tasks and collect loads of electroencephalography (EEG) data (basically collecting *brainwaves*, as electrical signals of the brain). Every day for three months, I put myself through long days of data collection, often falling asleep because it was so tiring to focus that intently for such extended periods of time.

At the end of those three months I tried training my AI designs, and what did they learn? Zip, nada, zilch! I spent the next month training, updating and retraining the networks. To no avail. Eventually I learnt more about better set-ups for EEG and how to check you have a clean connection resulting in cleaner data collection. It was then I realised I had been trying to train the AI with useless data – the artificial-colours-and-flavours-cupcake variety. I scrapped months of data and started again, making sure I was collecting quality data this time. Sure enough my programs started learning. After I trialled many different learning methods and ran it on a single computer for weeks, it eventually spat out a useful trained AI that had learnt to tell the difference between four different thought patterns to a high degree of accuracy (nearly 90 per cent). This allowed me to make my mind-controlled wheelchair a reality in 2008.

Back to big data. ML can be used to make sense of massive sets of data, ultimately finding potentially useful patterns and differences in that data. We ourselves are great pattern-recognising systems. Our brain even gives us hits of dopamine as a reward chemical when we find interesting patterns – contributing to why you feel good when you solve a puzzle, or if you see an actor in a movie and remember what else they've been in you tell everyone around you too. Just so they can get in on some of that sweet dopamine action – cos sharing is caring! With AI there are a range of approaches depending on what the AI system is trying to learn. Take deep learning, an example of ML that has flourished in recent years. Deep learning was a big step in the evolution of AI, allowing such things as a neural network program to receive masses of input data, find the patterns and differences on its own and extract the features that are likely relevant. Once it has learnt in this way, it can classify and contextualise new input data. It can even be designed to continue learning while in operation.

Let's use an analogy to understand the power of deep learning. Say you have a bunch of photos of animals from around the world and you give them to a young child. They would soon recognise patterns and could start to group the photos in categories of their own choosing. A deep learning system could be given a ton of these unlabelled photos and would not know what was what. But if trained to find patterns in the data through *unsupervised learning*, it would start to figure out patterns and differences on its own, creating various groups that the photos would fall into. It could recognise over time (and after seeing many, many photos) that if it saw a photo of a dog, it could bucket that into the collection of photos in which it has placed other dogs, without necessarily *knowing* what they are. Depending on how it learns, it might also throw wolves and other animals into this group. The child would likely do something similar, but then again some children might like to group the photos in terms of colour, so you can never be quite sure how this approach will turn out. The same goes for deep learning AI. The result may determine useful patterns in masses of data that may not have been apparent to the human eye or human-facilitated analyses.

On the other hand, if you were there to help the child learn through sharing your knowledge, you would show them a photo of a dog and let them know it's a dog. Over time they would recognise what a dog looks like and know one when they saw one. This approach is called *supervised learning*; with a deep learning AI system, the photos would have been labelled. In *semi-supervised learning*, only some of the photos would have been labelled by a human. This labelling allows the system to know what the answers are in those cases and then organise the rest of the data in the same way. This is how your smartphone and cloud photo applications can automatically curate albums and even allow you to search for specific things. If you search through digital photo albums to find when you wore that red hat, because the system has already been trained on many other photos online, it already knows what a red hat looks like, and so it can identify which of your photos include a red hat.

When rewards are involved, such as when a deep learning system is trying to learn a game, it often falls under the ML category of *reinforcement learning*. Desirable actions, such as collecting useful items and progressing in the game, are reinforced with rewards (usually programmed as points), while undesirable actions, such as losing lives or having to restart a game, are reinforced with penalties (usually deducted points). This is an approach that works well for games and simulations.

Deep neural networks basically have large sets of artificial neurons in layers, each of which can start looking for different patterns emerging from the range of data they have been fed. If the purpose is *image classification* to allow an AI to recognise objects like animals, and it has been shown a million pictures of various animals, and a portion of these had dogs in them, many of the neurons will structure themselves to look for various patterns recognisable in the dog photos it has seen. Many other neurons will structure themselves to look for other common features they've seen in the large image dataset, such as cats and cows and other animals.

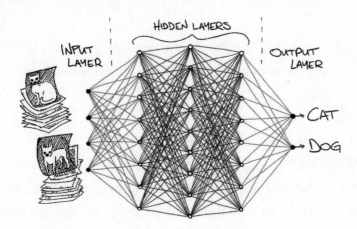

Figure 3: Deep neural networks

The *deep* part of the learning is down to the multiple layers of these neurons. Just say we had a deep neural network with three hidden layers. The first hidden layer might have neurons that look for tiny patterns in photos, such as edges of various shapes; neurons of the second hidden layer might each look for slightly bigger shapes, such as eyes and nostrils; and neurons of the third hidden layer might individually look for large shapes, such as entire figures of animals. If the trained deep learning system was to then see a new photo of a cat, various neurons would find the shapes they are looking for (particularly those looking for kitty shapes) in this new photo, and so they would activate. Based on the combinations of activated neurons, the system could safely say with a high probability that it's looking at a cat.

We do something similar within our own brain – though the processes of our brain are far more complex and capable than this approach – but here is something beautiful about what happens with us. A particular type of neuron we possess, known as a *mirror neuron*, form a physiological basis for empathy – our ability to understand and share the feelings of other people and to learn skills from them by watching. If you find yourself walking through a crowded place like a city, you may walk past hundreds or thousands of people, but if you see a single person look you in the eye and smile at you, you'll remember it. For the rest of the day. For the rest of the week. Possibly for the rest of the year. And maybe even for the rest of your life.

Why? Well among many other things your brain and body do in that moment in response to receiving the eye contact and the smile, many of your mirror neurons are firing and having a party as they are activated by that kind of interpersonal connection event. Even some of the neurons that fire when you smile yourself will be activated just by seeing someone else doing it. The level of feeling you receive can be indicative of how empathetic you are at that point in time. I say that point in time, because it's something that can change throughout our lives, meaning we can all work on becoming more empathetic. But isn't this an awesome thing? We have neurons that will activate when we see someone smile, and even more so when they're smiling at us.

I've witnessed my father work on many of these kinds of technologies over the years, and learnt all about neural networks from him. The RTX robot used these for a range of operations, and there was another machine set up nearby in that lab for experiments on computer vision. A computer was connected to a camera facing down onto a table with a stack of envelopes on it. You would take an envelope and handwrite a postcode, then place it under the camera. With a neural network trained to recognise the shapes of numbers, having learnt from a range of handwritten data, the computer would recognise the numbers written and display it on the old-school tube computer monitor. It would also show the probability for each handwritten digit across all ten digit possibilities, so we could see how certain it was of its classifications and if there was potential confusion with other digits. This process, called optical character recognition, is now used to convert a wide range of handwritten or printed text – from scanned documents, photos and the like – into digital text.

This was the beginning of the 1990s, and the field of AI has taken astronomical leaps since then. In 2015 I met Nolan Bushnell, one of the founding fathers of the video game industry, who in 1972 co-founded the pop-culture-icon company Atari (interestingly, the name is a reference to a check-like position in the game of Go). He was the first person the late Steve Jobs ever worked for. Nolan joined me on stage for a keynote I was giving on robotics and AI at a large creativity conference called Wired For Wonder – an amazing experience to share with a man who has had such a huge influence on computing and games of today. Backstage before we went on, we were sitting and having a chat. He took out his phone and asked me in a deep, American voice, 'Have you ever played *Crossy Road*?' I hadn't even heard of it at the time, but he showed me this 8-bit endless arcade-style, *Frogger*-esque mobile hopper game, where you navigate your little character of choice to hop across traffic or across logs over water, avoid trains and collect coins. He handed me the phone and as I played we talked about games of the past. Some of my favourite Atari games were the iconic *Space Invaders*, *Pong* and *Breakout*. Before founding Apple Inc., Steve Jobs and Steve Wozniak had both been involved in the development of *Breakout*, made to be a one-player version of *Pong*.

At that time I met Nolan, some significant recent events had seen AI programs by DeepMind Technologies (an AI company from London that had been acquired by Google) independently learn to play 49 different Atari 2600 video games. It learnt through deep reinforcement learning by playing the games many times over and simply figuring out from the pixels on the screen what it could control in the game, and over time working out which actions led to rewards and which led to penalties. It soon outperformed professional human players in more than half of these games, including *Space Invaders*, *Pong* and *Breakout*. What's more impressive is that it was often able to master these games within a matter of hours.

DeepMind soon went on to create AlphaGo, which toppled Lee Sedol in that historic match of 2016. This system first learnt from a dataset of more than 100,000 Go matches as a basis for its foray into the game. It then played many games against itself through self-play reinforcement learning (where it plays against itself). Move 37 of Game 2, the one that shook Lee Sedol, was significant because AlphaGo calculated that the chance of a human expert playing that move was one in 10,000. It was not a human move and at first looked like a mistake. It soon became apparent that it was strangely brilliant and different. To Lee Sedol, in retrospect, it was *creative* and *beautiful*. After five games over five days, AlphaGo emerged the victor, winning four games to one. The following year, now known as AlphaGo Master, it continued improving through playing many professionals. It went on to beat 60 professionals straight, including the world number one, Ke Jie, with three games to nil. It was another significant moment in the advancement of AI, but it didn't stop there.

Now that the world's best player in the ancient Chinese game of Go was this AI, the only true opponent would be a newer version of itself. This came in a next-generation version called AlphaGo Zero, which learnt Go in a different way. It was given the basic rules and then learnt from scratch, only playing games against itself through self-play reinforcement learning. It literally started with no prior knowledge of the game and used no human play in its learning. After three days it won 100 games to nil against the version of AlphaGo that defeated Lee Sedol. After 40 days of self-training it even convincingly surpassed the level of AlphaGo Master.

Another step forward and another newer design, AlphaZero, was created as a single system to learn from scratch and teach itself more than just Go. In just 24 hours of learning it beat not only AlphaGo Zero at Go, but also mastered chess and shogi (Japanese chess). This is thanks to its ability to play countless games and learn through experience so quickly that it can rack up much more game play than any human could possibly experience in a lifetime. That's the power of computing these days. These advancements continue and we will soon see many more mind-boggling capabilities of self-learning AI.

Along the road towards a general AI, these systems will become more adaptable, self-learning many things about the world beyond gaming. Over time, features like improved contextual understanding, connection, creativity and even emulated *self-awareness* and *curiosity* will evolve these technologies even further. It's not a particularly comforting thought for many, especially when we start to think about the changes the age of automation and AI will bring to the workforce. The age-old question remains: will the destruction of jobs outstrip the opportunities, or vice versa?

8

DOES AI NOW WANT YOUR JOB?

IT FOUND YOUR CAT IN A PHOTO SO NOW IT'S TIME TO ADAPT

All right, all right . . . so yes, AI can efficiently find cats in photos and destroy world champions in chess and Go. Does this lead to automation of jobs? Well, the short of it is, yes. But the other side is that new jobs are continually created as we humans *adapt* to the creation of new possibilities. There are many opinions and ongoing arguments on this topic, but what about long-term changes to humanity?

Well, firstly, history won't necessarily give us all the answers, because this time, with the exponential explosion and growth of disruptive technology, it may be different. If history is anything to go by, however, we've consistently been on the verge of being automated out of jobs for a few centuries now. Transitions to new ages in technology have sometimes been slow and painful, but our civilisation has continued to adjust, to the point where we are now having significant impacts on the resource output of our planet.

In 1890, much of the American transportation system revolved around the horse and cart, with thousands of companies building horse-drawn carriages, and there was a thriving industry around the raising, feeding, maintenance and cleaning of horses. In 1913, Henry Ford – one of the great innovators this world has seen – started a motor vehicle assembly line that over time charted a new course for transportation, leading to many new jobs, the growth of cities and an even bigger global industry. At the time, of course, these unknown emerging technologies would have caused great fear of job losses among those in the horse-based transportation industry. But who could have envisaged such technology would progress to create massive vibrant automotive industries on one hand, and eventually have a negative effect on the environment and contribute to climate change on the other hand? See, we often have a tendency to only see short-term loss or gain. Sometimes these issues are greater and more deeply impactful than we can ever foresee in the beginning.

This is not to say that AI won't take jobs, because it definitely will. But at the same time, new jobs – some previously unheard of and even not yet imagined – will arrive. What matters is how we deal with these oncoming waves of transition and transformation. While I was in China hosting the Discovery Channel series *Smart China: Start Up Revolution*, I saw first-hand how AI and agriculture drones the size of bikes are ushering in the future of farming and changing age-old traditions. Thousands of farmers had been dying each year as a result of cumulative inhalation of fertilisers and insecticides while manually spraying their crops. Now, farmers can put in an order via social media for their fields to be sprayed by the drones. Back at headquarters, an automated program calculates the size of the field, the minimum amount of pesticides or fertiliser spray required and where. A technician then arrives in a van to unload the drones and they take off to automatically cover the entire field. Surveillance drones had already been used to see what it all looks like from the air, finding where the roads, boundaries and obstacles are located. They can even identify if there are crop diseases in the fields and detect weeds or different types of plants that shouldn't be there. Humans work on computers, updating the AI on boundaries and obstacles from its extensive bird's-eye-view maps. So although the farmers here have had some of their own jobs automated out, they have more time for other, higher value things that need doing around their own farms, and will enjoy improved health and life expectancy. It also created a new style of digital work for those who teach the AI what's what, labelling masses of data (such as images with boundary lines) for the supervised level of learning.

I have seen AI used in some very polarising ways though. AI surveillance is an area that is rapidly evolving in capability. In Beijing I witnessed how cameras can be used with deep learning to recognise where people are, locating everyone even in large crowds. A sophisticated camera is able to monitor areas, using AI to locate the individuals within the feed. It zooms in and photographs people's faces in the crowd ten times a second! It can then identify individuals based on other collated photos of them, which means it can track everyone's movements.

Now the idea for all this was sparked by reports of international terror attacks where only a few security camera photos had been released of the suspects, and these were blurry and difficult to identify. This style of system can bank up what many individuals look like and track their movements. The data is never really intended to be seen by a human unless someone becomes a suspect in such an event as a terror attack, and their photo is added into the system as a suspect. The system will then locate where it has seen that person from these high-tech surveillance cameras and trace their movements back.

I tested out these systems, walking around for ten minutes with a group of ten people in a park watched by these cameras. We were all wearing winter coats, and I even put my hood on to hide from the surveillance cameras. When we went back to the office, the results showed that it had seen me and recognised me – after they had put me in the system as a person of interest. Despite my best efforts, I had not hidden at all. The cameras had even snapped shots of me from 25 metres away, tracking nearly all of my movements around the park.

A real Big Brother kind of system this is, and definitely a controversial and polarising one. I personally feel very uneasy about such surveillance technology. The way I view these advancements is that they can spark new positive possibilities in addition to the dark sides that are quite easy to identify. A surveillance system such as this could potentially achieve what it was designed for, to recognise and track suspects or people of interest to authorities when needed. It will further evolve into an ability to prevent disasters, though these systems will be rolled out in some countries and rejected by others. Privacy and surveillance laws always vary from country to country, but do we really want such systems monitoring our every move? It should be noted these systems use machines to store, process and operate data, not human surveillance. That may ease the minds of some, but for many, freedom will outweigh the benefits from constantly being under surveillance. We should always weigh up both sides of emerging technologies, the positives and the negatives, and decisions should be made about what really matters to the people. Knowing about the technologies to come and raising important conversations is the first step we can take towards shaping future changes.

AI and camera technology can do wonders in nature conservation. I witnessed one such application at the start of 2020 while filming a documentary in a large nature reserve. A scientist and I were out in the snow, tracking the endangered Siberian tiger (also known as the Amur tiger) and the Amur leopard. We were merely a few hours behind a large female tiger who had passed through, leaving huge paw marks as she went. While we set up and tested some camera equipment, I learnt about what they do there. In a grand-scale protection and conservation effort (with World Wildlife Fund International and Intel collaborations) thousands of 'camera traps' (which are triggered by movement) are being mounted to trees in planned locations to record and track activity in this massive nature reserve.

Deep learning AI programs automatically detect which animals are in these recordings and catalogue them through the cloud to monitoring stations. It will do this only if animals are in view (discarding falsely triggered footage), allowing for the tracking of movement patterns, endangered species numbers, types and numbers of prey, and amazingly being trained to recognise individual tigers from their unique stripes (like identifying a human by their fingerprint). In addition, these systems can alert rangers if poachers are detected. With AI helping to do much of this identifying, sifting, sorting and cataloguing work, the researchers can get the data they need faster and focus on other tasks. The more the researchers know about the endangered species and where they are, the better they can protect them. I hope that even the need for these types of projects can be a wake-up call. Preventing such species going endangered in the first place and protecting our wildlife is critical. But now, this type of application for AI is one of the most inspiring I've seen and I'm certain we'll see many more similar conservation efforts for wildlife and nature.

This style of approach can be beneficial in other ways, too, such as detecting if elderly people in their homes or in high-care facilities are lost, have fallen or need help. Human monitoring and assistance will likely not keep up with the growth of global ageing populations, so technology may have a big role to play here. It could be specifically designed to improve safety and independence for the individual, rather than track their every move for any other reason. With understanding and transparency, we can as individuals, as collectives and communities, as societies, steer these constantly advancing technologies.

Individual examples like these showcase how quickly things can change. And given that AI can be activated and respond much faster than deploying humans could, not only will some jobs go, but new ones will also sometimes go straight to the machines. The truth is, over time, all applications, jobs, industries and sectors will be affected one way or another. That's just part of intelligent advancement. I will stress here, although it has been estimated many times over in countless studies that a large range of jobs will be at risk of being overtaken or at least altered by the likes of AI, more new jobs for humans will be created in their place. Roles that involve a lot of repetition and routine are most at risk – but this offers a growing opportunity towards creative roles, complex roles and those that involve compassion, positive social and environmental impact and social interaction – and we all know how humans generally crave human interaction.

It all comes down to how we deal with these changes, recognising that what we consider to be soft skills will be increasingly useful and important in helping us adapt, connect creatively and shape the many changes coming our way. Stay curious and keep learning new and interesting skills to aid in adaptation. Like chameleons. They don't just wait because they can't be bothered changing with the environment. They recognise change directly around them and adapt their colours accordingly, for camouflage, to communicate, and to reflect and absorb heat to regulate body temperature. I'm not saying you need to blend in. I'm just saying adapt. Like a human-chameleon of transferable interests.

If the world is changing, let's continue creating new careers to change it for the better. Together we can build our own brave new world.

If AI and automation bring about unprecedented change,
we must face it head-on.
We will dream, adapt, imagine, create,
to harness these powerful technologies for
the betterment of humanity,
and to shift our focus more towards our
interpersonal connection,
our empathy,
our human experience.

9

TO HUMANITY OR NOT
TO HUMANITY?

THAT IS THE QUESTION

We've seen already that AI comes with a raft of terminology for the methods it uses to learn. More broadly, the various levels of AI fall into three common *theoretical* tiers with reference to human intelligence:

1. **Artificial narrow intelligence (ANI)**
   AI is narrow in its abilities and is limited to learning in specific areas, but it can still perform those limited abilities incredibly well.
2. **Artificial general intelligence (AGI)**
   AI can draw upon its varied learnings and effectively reach human-level intelligence.

3. Artificial super intelligence (ASI)

AI goes beyond general intelligence, and thus beyond human intelligence, and exponentially improves on its own into a form that far surpasses the intellect of our species.

These theoretical AI tiers are anthropomorphised here with a robot illustration. Let's examine each in more detail.

Figure 4: Theoretical tiers of artificial intelligence

ARTIFICIAL NARROW INTELLIGENCE (ANI)

This basic form of AI is also known as *weak AI*, but don't be fooled – ANI has been and will continue to be used for some mind-melting, groundbreaking things. It can learn very narrowly defined tasks, up to and exceeding the capabilities of humans, usually with a view to making big steps in a single subset of abilities. ANI covers everything we've achieved in AI so far – mind-controlled smart wheelchairs, driverless vehicles; healthcare and triage, utilising data of the past to recognise potential diagnoses; prediction machines; extracting intention and context from natural language; learning to speak a language like chatbots and even translating between languages; machine vision detecting people and remembering them; and excelling at games such as chess or *Jeopardy!* or Go.

The reason these are all *narrow* in relation to humans is that even AlphaGo, with the incredible abilities it used to topple Lee Sedol in that groundbreaking five-game match, would have no idea how to read you a children's book. And that's because it was made to excel in a single subset of abilities, in this case the patterns and strategies in the game of Go. This is the tier we're currently in, and all AI designs thus far fall under ANI. As our implementations improve, diversify, connect and converge, we will take critical steps towards the next tier, artificial general intelligence.

ARTIFICIAL GENERAL INTELLIGENCE (AGI)

Also known as *strong AI*, this is human-level AI. For humans to achieve a general level of intelligence, they must be able to draw effectively on numerous varied abilities. In AGI this will be done to such a great level that programs will successfully operate on par with the human brain and intellect and, in effect, exhibit the full range of human cognitive abilities. This AI will show intellect, logic, intuition, adaptability, creativity, will beat you in Go *while* reading you a children's book, and may even display emotions and empathy.

Empathy was what threw off the robots known as Replicants in *Blade Runner*. In one great scene an interrogator closely watches the response of a Replicant being questioned about a hypothetical situation that for humans would elicit an emotional and empathetic response. This is a take on the Turing test. The physical appearance of the humanoid is so lifelike to the human eye it is indistinguishable from a human, and its reasoning and answers are enough to trick most humans, hence the need for deep questioning. True AGI, however, will undoubtedly pass Turing test variations, no matter what level of questioning you throw at it. This tier is unlikely to last long, as an AGI would very likely be capable of improving upon itself quickly – moving towards the next tier, the ultimate tier, the tier that shatters all tiers ever in the history of tiers. Artificial super intelligence.

ARTIFICIAL SUPER INTELLIGENCE (ASI)

This is it. We're now literally talking about the invention to rule all human inventions. Ever. Period. To reach this tier we will have passed the hypothetical point in time where AGI has evolved an ability to increase its own intelligence exponentially, far beyond human intelligence, including the brightest minds of our species. We call this the *technological singularity*. The point of no return. In other words, put on a few extra sets of those undies you're wearing, because if things don't get really, really good for us at this point, they'll get really, really bad. One way or another, if ASI is made possible, the changes to humanity and civilisation as we know them will be so extreme it's difficult to fathom. What we do know is that it will be *uncontrollable* and *irreversible*. We will have created a new alpha 'species'.

We will no longer be able to simply pull the plug, find the kill switch or revert back to the information dark ages by taking out all satellites, servers and computing systems, the backbone of AI. None of this is a possibility, unless the ASI lets us – remember, by this stage it is not only smarter than our now primitive selves but it also has control of all connected technology – defence systems, vehicles, weapons, computers, rocket control systems, robots. We're not going to have much left in our arsenal.

Humanity as we know it will completely transform in one way or another. We will have built god-like entities with intelligence, reasoning and problem-solving capabilities we can only dream of. If the values and goals of such an ASI are not aligned with our own, we will be merely an anthill in the way of whatever it aims to achieve. But if we continue with these important conversations and build in alignment between AI and humanity, it may take us to seemingly infinite new possibilities, helping us to really understand the universe and even find ways to traverse it, to live in sustainable harmony with our planet and our solar system, while also being set up with an ability to escape the confines of our earth with ease. If a utopian world is what we want, ASI could give us the blueprints to start creating. If quality immortality full of adventure is what we desire, ASI could show us such a possibility. If reaching the stars is what we seek, ASI could design the transportation or teleportation methods necessary. We would still be limited by resources, but we will understand how to harness more of what we have, while potentially requiring much less to operate – and so humanity could become more efficient. We may even merge with technology to help achieve this, becoming cyborg and bionic.

Of course, many problems seem to outstrip any theorised positives of this level of AI. Humans would likely be much more void of purpose if everything could be done and created by ASI, and resting on the hope that it would choose to help us is foolish. Humanity must actively shape the possibilities to come, whether we want to prevent elements of such technology or guide its advancement towards those beneficial for our existence.

For now, we are firmly placed in the weak ANI tier, moving rapidly towards strong AGI. Having said that, it's difficult to know exactly where that point is, and truly definitive tests for AGI would need to be carefully devised and well-rounded to even be certain we've reached it. We could find an AGI that passes the Turing test with flying colours when interviewed by one human subject, while failing dismally with another. That's because human interests, intelligence and capabilities are astoundingly vast and varied. And the truth is, AI can be built to simulate human emotions or intellect well enough to make people believe it might be conscious. That doesn't mean it is.

This raises another profound philosophical dilemma. How do you know with certainty that any person other than yourself is conscious? You know you are living, that you experience the world around you, that you are human and that you are self-aware. How can you be certain anyone else you know is too and not just seemingly so? You can *know* it to be true, but you can't yet *prove* it.

This kind of dilemma, only one where we'll be more sceptical about the outcome, will start to occur with AI *before* we hit true human-level AGI. I predict we'll see an *artificial consciousness paradox*, in which AI will be able to sufficiently dupe humans into believing, thinking and feeling that the AI experiences subjectively and lives with consciousness. It has the potential to effectively emulate all manners of feelings: fear, sadness, excitement, empathy, even love . . . without actually consciously experiencing them. But it will definitely be difficult to tell, and then we'll have a similar dilemma to the one described above for humans. You can *know* these machines don't actually feel, but it may be very difficult to *prove* it. So then what happens? Do we give these systems rights? Will they feel the need to have rights? We may find ourselves searching for tests like the one from the *Blade Runner* Replicant interrogation, searching for the presence of those small human nuances when put through emotional and empathetic reasoning – only AI doesn't necessarily come in humanoid form, so this will definitely call for an advanced form of the Turing test.

One thing we do know is that AI is here to stay, no matter what forms it takes on in the future. A school of thought exists that we're moving into an age where we ourselves will merge with AI, and to some extent even connect our minds to the internet. Either way, we're consistently outsourcing more and more of our own cognitive processes to technology. If a greater level of intelligence were available to us on demand, how might that change your life and what would you go on to do with it?

An important factor is that even ANI designs may help us lead longer and better lives. AI that tracks a range of our personal data, including the kinds of health tech that are already on the rise, can monitor all the stuff that's going on with us – depending on how much data we're willing to give it to perform this duty.

I've seen first-hand how AI can merge with health tech, DNA profiling and social media to provide powerful insights into our own health and our probability of ending up with various diseases and conditions.

I met one of the young co-founders of a company who used a form of *virtual twinning* technology (we'll explore this further in Part V) to have a curated diet created for him to manage his AI-predicted predisposition to diabetes. He has not yet been diagnosed with this condition, and hopes he never will now he's been given this powerful data-driven information. Insights like these will improve as more people enter such platforms and provide a greater range of base data for comparison.

Now this may seem quite confronting and a little unnerving in terms of data privacy. Many of us are not instantly keen to give away that amount of personal information about our lives. But what if the outcome could guarantee you an extra twenty quality years of life? Now how protective do you feel about that data?

This is all taking steps down the path towards a sci-fi–style period known as the *posthuman* era, but to get there we're steadily moving into a transitional period, the *transhuman* era. This is the idea that we increasingly integrate our biology with technology, until we become one over time. These philosophical eras lead us towards redefining what it means to be *human* and charting our own next evolutionary steps. If we move so significantly past our current understanding and form of human – say, if we were to somehow completely digitise or quantise ourselves into immortal beings, transitioning our consciousness entirely outside of our own biology – we will by that point become *posthumans*. This is largely imaginative, philosophical, and very sci-fi as far as abstract concepts go. But given the advancements we've already witnessed, combined with exponential growth in technology and AI . . . well, nothing is impossible. I'm not telling you how to feel about it, I'm merely mentioning these as topics that could very well be relevant to our future humanity. I'll leave it up to you to decide how you feel.

The truth is, AI is one of the many building blocks of technology. So will this new super force be one that leads to destruction and annihilation, or one that helps humanity thrive into the future, finding ways to fix our world and maybe even allowing us to reach the stars?

It's a bold leap of the imagination into the future, trying to predict what this type of technology could be capable of. The ability of intelligent systems to assist in our understanding of where the future is headed is one of the great benefits AI provides. In my 2017 trip to Tibet, filming *Living on the Roof of the World* for Discovery Channel, I saw one astounding use of predictive analytics, using lots of data with the likes of statistics and AI to make predictions about the future, in research on the Tibetan glaciers. I trekked up to the glaciers with a team of scientists, our film crew, local Sherpas and guides. After a rough four hours, we finally reached the base of this one glacial site and, with chain-saws and tools, the scientists take samples. This has been done for a number of years at a range of sites, resulting in a massive storage bank of glacial ice, labelled with the various locations from which they were acquired. These are used to analyse how old the ice is at each site and determine whether the glaciers are increasing, stable or declining.

The way they calculate the age of the ice is to carefully drill into each sample, then extract and radiocarbon date the likes of plant matter held within. Why? Well, the older the ice on the top of the glacier, the more frequently ice has been melted off the top due to such factors as climate change. I even held a sample of ice core that had been dated at roughly 20,000 years old. Over time, these scientists have gathered masses of data on each of the Tibetan glacier sites. This large-scale project is very important because more than 46,000 glaciers on the Tibetan Plateau supply nearly 2 billion people with water. And they give birth to rivers that flow through China, India, Pakistan, Bangladesh, Burma, Laos, Thailand, Vietnam and Cambodia. These include such important waterways as the Yangtze, Mekong, Salween, Indus, Brahmaputra and Yellow rivers.

Patterns can be found in the data, and through predictive analysis, computer models can be created to make future projections and better understand the impact of the glacial retreat. If humanity harnesses its own intelligence alongside that of our artificial computing counterparts, we may mitigate great risk and even conflict. The analysis so far has found that the glaciers have been in major decline since the 1950s and that the ongoing affects could be catastrophic. Over the past 50 years, the temperature in this region has increased by about 1.3 degrees Celsius, three times the global average, and if this trend continues, it is believed that 40 per cent of the plateau's glaciers could disappear by 2050, meaning these countries need to seriously consider alternative water sources before then.

In the past, wars have been fought over land and energy, but the next may happen over this dwindling resource as we see tensions rise, particularly between China and India. But intelligent analytics and predictions could possibly help find new solutions as well. Armed with this incredibly valuable knowledge, we could perhaps bring about much needed change to help alleviate the rate of decline, maybe even finding the major specific sources of temperature increase – one of which here has been found to be black carbon, an atmospheric pollutant. These very small particles are formed from the combustion of fossil fuels, biofuel and biomass, then slowly fall from the air and settle on the glacial ice. This soot, as it's also known, absorbs solar radiation and contributes to heating and melting of the glaciers. The more we understand our impact on our planet, the faster we can act to prevent further disastrous repercussions.

Where AI may play a greater role as we move into the future is in looking further than individual examples and bringing them together more holistically to examine complex sets of flow-on effects from single decisions. It may help us gain a glimpse of the future through data, and assist in solving the problems it identifies. The fear with AI comes from the idea that as it gets more *intelligent* it will also become more cognitive and self-aware. My take on the advancement of AI is that there are many different types all advancing at different rates, and when compared to our own brain, these types will not hit ASI level or even AGI level at the same time.

So how do *common sense* and *consciousness* fare in the artificial domains? Well these have not advanced as rapidly because, for many creators, their return benefit is simply not as easy to see. Humans can find immediate use cases for AI when it is still dependent upon our vision, our data and our tasks. We like knowing that we can *contain* such a technology and retain control over it. But in the wrong hands, the technology could be used by humans to assert control over populations of humans, and hence it becomes a new arms race. This is why we must think of ourselves as a single species and learn to thrive together. It has already happened in part. I've noticed the term 'human' used much more frequently in recent years, and I believe this is due to the perception of a growing, external *threat* – automation and AI. At the same time, we have also shown many signs of becoming more separatist across the world. Let's not forget our humanity, our shared experience as a single species inhabiting this one planet we all call home.

Instead of thinking of AI as a threat, if we see it as a powerful tool, perhaps even something of a human-made prophet fuelled by data, then we could use it to plan long-term ways for humanity to flourish. It could assist us in unravelling ways to decode our own immunity and become more resistant to disease, an endeavour made ever more pressing given the devastation of the 2020 COVID-19 pandemic. If this were the case, and humans lived longer, healthier lives, AI could also be used to project the impact longer-living humans will have on the earth. Will short-term life extension lead to long-term disaster for our species, or will there be further flow-on benefits? Instead of *fear*, let's continue cultivating *hope* and a desire to understand and act towards a better world.

When I was young I remember hearing about how Earth had a hole growing in the ozone layer, and that without the ozone layer we were sitting ducks for an unprotected onslaught of the sun's relentless radiation. And we humans were causing that hole with something called CFCs, or chlorofluorocarbons, chemicals used in aerosol sprays, air-conditioners and refrigerators. The Montreal Protocol, an international treaty designed to protect our ozone layer, came about in 1987 to phase out the production and use of many of the responsible substances, starting with CFCs. As a result of this international agreement, and its widespread adoption and implementation, the hole in the ozone layer over Antarctica has been slowly recovering. These agreements take great vision, clarity and international collaboration to begin in the first place, and then to be upheld over the many years of implementation. The inspiring thing is, as this example shows, humanity is capable of it. It's not the only time important international agreements with global reach have occurred, and I hope we'll see many more – for the benefit of our one planet and the life it harbours.

If we do indeed reach the widely acknowledged idea of technological singularity, where intelligent systems move beyond humans in every way possible, including consciousness, then thinking we can retain *control* over it is ridiculous. Imagine giving chimpanzees, one of our species' closest living relatives, a chance to create a cage that a group of humans cannot break out of. It's inconceivable that they could succeed, because our intelligence and ability to solve problems will go well beyond any ideas and implementations of a cage chimps could come up with. Similarly, there is no chance we could ever cage or control an AI that has surpassed our own human intelligence. Instead of attempting to cage it, if we roll back development ahead of time in some areas of potential AGI intelligence – such as self-awareness, self-preservation and consciousness of existence – then what we may be left with is something amazing. We will need to have built in positive human value systems, so that the goals of AI are aligned with grand humanitarian visions of harmony between life and Earth and beyond. We may then have a humanity-preserving super-powerful solution-creating machine!

Think about it, the ability of computers to perform calculations combining complex mathematics, large numbers, calculus, arithmetic and statistics already goes well beyond human capability, so in essence some elements of AI have already well surpassed humans. Our computers didn't suddenly get up and try to wipe out humans because they calculated that we are inferior mathematicians. In fact, they have no comprehension or understanding of mathematics, because cognitive abilities have not been built into them. As a result, humans can harness our computers to run calculations we cannot do ourselves. They are a tool we use towards achieving our dreams.

In a similar way, we could build amazing problem-solving AI that could examine many possible solutions and innovations humans could not otherwise conceive of, and even predict the flow-on positives and negatives of each solution. If this were to be made in a similar way that we can ask a machine to master a game like Go without it having any real comprehension of what it all *means* or a true understanding of its own existence and the world around it, then this will undoubtedly become a useful invention for our species. Since realising the importance of such technology I've been set on a path to contribute to steering the directions we take with AI and helping others to understand and shape its impacts.

The funny thing is, we haven't yet figured out exactly what *consciousness* is, but it's clearly significantly more complicated than inputs, processing, calculations and outputs. We are special types of beings. Evolved to adapt, to harness the elements, to create tools, to communicate, to collaborate. If we cannot figure out what consciousness is, how could we possibly create an artificial version of it? What I find is, the more we advance our technologies, the more we understand ourselves – through new tools to measure and analyse, and through general reflection upon each advancement. Ultimately, we're always learning more about what it means to be human.

We do, however, find it more difficult to imagine the impact these agents could have if freed of our shackles to create their own goals and strategies, even innovating through self-critique and criticism. This is the idea of *transcendence*. The hope and aim for AI is that we instil good morals, ethics and values before ever getting to a point where we could lose control. This is something we must never stop planning and preparing for, making sure people have a say on what gets coded into these machines as an underlying *purpose* system – an approach acknowledging Asimov's laws, but perhaps more quantifiable.

In contrast to the predictions of doom and gloom, we could utilise AI as a tool to help humanity thrive into the future. I believe we don't need to worry too much about getting AI to even emulate consciousness, but instead harness the technology for prediction, automation, efficiency, sustainability, disaster recovery and even prevention, problem-solving and solution-creating capabilities for the benefit of humanity. If we continue to advance AI, it needs to work within the value system we design. This may mean designing AI without striving to achieve artificial consciousness and without building in the capacity, desire or need for AI to go against its in-built rules and boundaries. We should also look to prevent AI making decisions and acting on taking human life without any accountability. These are starting points that will ensure we don't lose control of these creations.

If AI can augment us beyond our capabilities, this might be the sweet spot. In combination with bionics, it may enhance our inherent unexplored abilities. It may gift us ways of reaching the stars and exploring the far reaches of our universe. And given our current level of outsourcing of our own cognitive abilities to the likes of mildly intelligent smartphone apps and home assistants, our ability to adapt to these big changes could mean AI increasingly becomes a part of us.

A cognitive extension.
We may become one and the same.

The big thing to realise is that creativity, innovation and human experience will flourish in these times of AI and automation. So in this age of rising automation, we should remember the wonderful things about being human. Our conscious experience that breathes meaning and appreciation into our world. Without it, this incredible planet, the wondrous night sky full of stars and life itself, wouldn't be appreciated at all. These powerful technological tools should only be used to improve our humanity, and that is what we must strive towards. Your brain and body are an interconnected superhighway, your mind a natural-built supercomputer, your life a near completely impossible reality, your ability to love a truly magical gift and your consciousness a thing so unique we don't know of anything like it in the entire universe. And that, my friend, is a beautiful thing.

Transcendence may create the next great evolution.
Will AI transcend its boundaries of current inability,
or will we give it the ability to comprehend?
Will we humans transcend our biological limitations
and finite lifespans before AI takes over?
Will we instil purpose, morals, ethics, humanity?
Or will we, perhaps, become one,
and transcend together?

PART III
BECOME CYBORG AND BIONIC

Because who wouldn't want to be superhuman?

10

WE RUN ON BATTERIES AND ELECTRICITY

OUR EXTRAORDINARY HUMAN BODY AND BRAIN

I find working on electronics and robotics a great source of mental stimulation, but learning about the human brain and body is just so full of wonder – they are some seriously complex and incredible systems. They make me question things I had taken for granted or not even given a great deal of thought when I was younger. Things like that we are the product of many ages of evolution, that we can move our entire body without questioning how and that seeing is believing.

I distinctly remember first learning about the heart. It's such an intricate organ, far from perfect, but its inner workings are amazing. The pumping of the heart creates a smooth-flowing mechanism that moves blood from the upper chambers (atria) via the lungs (into the left atria) and body (into the right atria) to the lower chambers (ventricles), which then pump blood back out to the lungs and body. Valves like little doors slam shut to prevent the backflow of blood from ventricles back to atria, and from arteries back to ventricles. The left side pushes oxygenated blood around the entire body from the left ventricle, whereas the right ventricle pushes deoxygenated blood through the pulmonary system (the lungs). As blood travels through the lungs, it releases carbon dioxide (to be exhaled out through the breath) and picks up oxygen (from our inhaled breath), returning this oxygenated blood to the left side of the heart so it can be distribute to the rest of the body again, and the cycle continues. It makes sense, but learning how it all works is still something else.

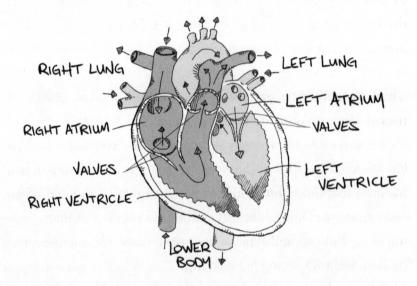

Figure 5: Our human heart

I was also stunned to realise that the heart operates on a flow of electrical activity, and that's what we see when we look at electrocardiography (ECG) signals from the heart. The beat all starts with the pacemaker. This is an autonomic (involuntary, acting automatically to keep us alive) nerve that sends electrical signals down through the middle of the heart, making it contract as it goes. It gets to the base and flows back up the sides, giving the heart its beautiful rhythmic beat. It does this rather than beating all at once, which actually wouldn't work because the atria and ventricles need to take turns pumping to control the flow of blood.

In 2009 I attended a conference in Minneapolis in the US, and the keynote speaker was Earl Bakken, the man attributed with inventing the world's first wearable, battery-powered pacemaker biomedical device. A pacemaker device can be implanted to continue sending electrical impulses that maintain normal rhythm in damaged hearts. Despite it being a cautionary tale of horror, he had been inspired by *Frankenstein* from his youth, the story of a scientist, Victor Frankenstein, who animates a monster with electricity. Bakken came to the realisation that 'when electricity flows, we're alive. When it doesn't, we're dead'.

I'll never forget the profound words he delivered at the end of his speech: 'I'm glad I built the pacemaker device and co-founded Medtronic to create medical devices for the people. Because one of our pacemakers and two other medical device types are implanted in me and have been keeping me alive for the past decade. I wouldn't be sitting here if I hadn't chosen to work on medical technology.' Jaw-dropping. He passed away in 2018 at the age of 94, but only lived so long thanks to his own creations. Creations that have also extended millions of other lives globally.

Much of this ability to fuse technology with ourselves, human and machine, comes from the field of biomedical engineering.

That we can do it at all relies on the fact that we have more in common with technology than most of us realise. It blew my mind as I learnt that, in a way not too dissimilar from our own computers, we also send data around the body in a massive, vibrant, connected information superhighway. We send electro-chemical signals (nerve impulses) from every one of our sensors through to our brain to process, and it simultaneously sends many of these signals across itself via the brain cells (neurons), to create thought, pattern recognition, intelligence, awareness and the seemingly infinite array of human characteristics, while also giving us physical control over our own body.

If we can already connect our bodies with technology, there must be many untapped fields of integration. Surely over time we will increasingly interweave human and machine. We used to call them cyborgs. We often still don't feel like cyborgs really exist today, but that's because we're accepting as humans of what we already have, so we look to the next thing around the corner. We've moved the goalposts. If we were to take someone who owns a mechatronic bionic limb just half a century into the past, the people of that time would likely believe they're witnessing a cyborg.

What do the terms 'cyborg' and 'bionic' actually mean? A cyborg (cybernetic organism) is an organic being, such as a human, that has restored or even enhanced abilities through the integration of technological body parts (often bionic). Bionics is the application of engineering in a biologically inspired way, made to re-create functionality from nature, since nature has an immense database of working solutions. Today we have cardio-vascular implants such as the cardiac pacemaker, sensory implants like the cochlear implant for hearing, neurological implants for stimulating or measuring nervous system activity and bionic limbs with feedback of touch and sensitivity. We're also making progress on the bionic eye. Basically, there is a constantly expanding range of technologies that can be implanted, connected or even fused with the human brain and body.

And what about our batteries? Well, as a biological system, we convert and store energy from the things we consume. Carbohydrates, protein and fat form our three sources of energy. The body's go-to energy source is carbohydrates. These break down into glucose, which is an immediate source of energy, especially for the brain and muscles. The energy we don't use we set aside for when we might need it later, converting any leftover glucose to fat. Our fat cells are our long-term battery storage. Our energy powers such things as our minds and imagination, our movements and our regulation of heat throughout the body.

In contrast, electronics require power supplies that feed them a flow of electric charge . . . electricity! But energy can be converted from one form to another (as the first law of thermodynamics states). Bionic devices, particularly those that can be implanted, always require a power source, and *we as humans* will increasingly become that source. Not in a slave-to-the-Matrix kind of way. I'm referring to us being a possible biological power source for small electronics. This can come from movement (our kinetic energy); our generation of heat, which can be turned into energy by small thermoelectric generators; and the harvesting of readily available ions (or charged molecules) from the body's fluids. Among other emerging creative ideas, these methods can be used to power such devices as low-power implantable biomedical systems. Refining these energy-conversion technologies will see more biomedical technologies powered by the human body – another step into the future of fusing human and machine.

Another interesting thing about the human body is that so much of how we operate is facilitated by electrical impulses. We humans have developed methods to collect and analyse these signals to learn more about the brain and body and to diagnose or treat illnesses. Apart from measuring the electrical signals of the heart with ECG, we can also examine the electrical signals in the muscles with electromyography (EMG), the electrical signals of the eyes with electrooculography (EOG), and the electrical signals in the brain with electroencephalography (EEG). Our biological signals can be used for analysis, and in the process, converted into digitised signals that can be displayed. But they also hold interesting clues into what the owner of the signals is *doing* or *thinking*. And if we can analyse something, that also means we can harness it. This is just what Bakken did when he created the pacemaker.

As much as we can break down our make-up, analyse it and understand it, we're still on a constant journey of understanding what it means to be human. Yuval Noah Harari says that factors making us an alpha species on this planet are our intelligence, our ability to both use and make tools and our ability to collaborate at scale. But why are we as self-aware and as complex as we are? Creatures and organisms in nature tend to evolve defence mechanisms and abilities for survival. These don't often go so far beyond the point that they attempt to conquer every corner of the globe. Humans, however, did just that. We're not optimised to surviving one on one in the wild against lions and bears, nor do we have a chance on our own in the water against sharks. What we did was learn to *collaborate* and *create*. And then, over time, we also developed our own forms of *competition* among ourselves, pushing the limits of our various abilities further and further.

The thing that intrigues me is the fact that we are much more complex, much more capable and have much more potential than we require simply for temporary *survival*. For better or worse, we choose to advance and to my mind, this process of advancement includes bionics. We have self-awareness and an ability to reflect on the past, analyse the present and imagine the future. Our capacity for foresight is unique and something special. It's what allows us to create technologies that can help us all. To the intelligence of human beings, survival doesn't necessarily mean survival today. It can also mean survival tomorrow. And the day after. And the day after that.

We live longer lives due to modern medicine
and technology.
We tackle challenges faced by the human body and brain.
We are learning to play God with ourselves.
And we are the architects of our own next steps
in human evolution.

11

BUILT BY NATURE

APPLYING NATURAL BIOLOGY
TO HUMAN TECHNOLOGY

It's 2007 and I have finally made it to working on smart wheelchairs for my undergraduate thesis and my PhD. I work on two different robot wheelchairs, the first named SAM (for Semi-Autonomous Machine) for half a year; and the second, more advanced robot wheelchair, TIM (for Thought-controlled Intelligent Machine) for about five years. SAM gives me my first insights into the many challenges that come with taking control of a wheelchair through a computer. The chair is connected to my big tube computer monitor on my work desk in the Centre for Health Technologies at UTS. I'm trialling design after design on the code to provide my programs the ability to control the wheel motors and take control of SAM. But this proves much more difficult than anticipated. Each time I attempt a new test with my hand on the emergency stop button, nothing happens. So I grab some lunch, throw the code into *debug mode* so I can find what's going wrong and where, and munch into my sandwich as I lean back in my seat.

Little do I know that *debug mode* will actually get rid of the problem preventing my wheelchair from running, known as *interrupts*. Basically, my code isn't given enough time to get all the way through its calculations and send the result to the motors before it starts again. Now, with this barrier removed, and me leaning back in my chair completely unaware of it, SAM suddenly takes off, pulling the monitor off the table. I lunge for the screen, catching it before it smashes on the ground and immediately look up towards the runaway chair . . .

BAM!

Straight into a wall smashing a hole in it and bending both footrests. *Agh! I'm going to need to get those repaired*, I think. *On the plus side, it works!* Time to give it an ability to *see* so that it might avoid these types of accidents.

The more I make my way through the world of technology, coming up with ideas and designing, the more I start to realise the undeniable influence nature has had on various technologies of today. From the sensors that allow robotics to perceive their environments – in other words, converting real-life information into digital representation – to their mechanical physics and movements and the intelligent algorithms operating their abilities, much has been gained from nature. Having had millions if not billions of years to get things right, nature provides a wealth of inspiration for many facets of technological innovation.

Mathematical principles similar to those used in robotics applications are extremely useful, and in some cases necessary, for a more complete understanding of the human body and how we interact with our environment and each other. They can also allow us to better learn from nature and continue to improve our own technologies. Our sensory technology is modelled on things in nature, such as our own senses, as well as those we discover in the animal kingdom, including unique ways other creatures perceive their environment. But instead of feeding the data as electrochemical signals to the brain, which is how our own systems operate, technological sensory systems usually take the information they sense from the real world and convert them to digital streams of ones and zeros to be interpreted by such systems as the most common form of computing 'brains', central processing units (CPUs) and graphical processing units (GPUs).

We use cameras practically every day now that we all have them so conveniently built into the devices we take everywhere with us. Originally designed to capture moments in light, cameras have undergone much evolution over the generations, as have our knowledge and understanding of eyes and vision systems in nature. The more we've learnt about our eyes and those of other living creatures, the more the abilities of cameras have been able to advance. Take stereoscopic cameras, which are modelled on the human eye. We have binocular vision, meaning we can see everything from two slightly different perspectives. We don't have two eyes just because it's more aesthetically pleasing than having one (though if we were all Cyclopes we might think two eyes were strange), nor do we have both as a redundant backup system in case we lose one. Our two eyes actually help provide us with three-dimensional (3D) vision of the world.

You may have seen the *Simpsons* Halloween episode 'Treehouse of Horror VI' where Homer gets stuck in a 3D virtual land, and this concept is described so perfectly by the scientist, Professor Frink. In his overly geeky voice he draws on a black-board and says, 'Here is an ordinary square.' Policeman Chief Wiggum responds, 'Whoa, whoa, slow down, egghead!'

'But suppose we *extend* the square beyond the two dimensions of our universe,' the boffin continues, 'along the hypothetical z-axis there. This forms a three-dimensional object known as a *cube . . .*'

I love this. As our universe isn't a 2D cartoon, the z-axis is a very real one for us defined in the Cartesian coordinate system (another inventive contribution by René Descartes), allowing our reality to exist in three dimensions: the x-axis, the y-axis and the z-axis. So how do we (and some robots) see in 3D? Each of our two eyes provides something closer to a 2D image, but as they bring in two slightly shifted views of our environment simultane-ously – at exactly the same time – they help us see in 3D through a concept known as *parallax*. To see this process in action put your index finger up in front of your nose (at the distance you would hold a drink in between sips). Close one eye, then switch which eye is closed. You'll notice a change, as if the finger moves position when you switch eyes. Alternatively, with both eyes open, focus on any object beyond your finger and you should instantly notice two slightly shifted ghostly versions of the same finger. This is *parallax*. It's caused by our different viewpoints of the same object and is an important principle in understanding 3D vision in animals and other creatures, and it also plays a huge role in the way many technologies (including robotics) perceive their environments. Your brain also has ways to figure out how far away an object is. One of these is by comparing the distance between the two perceived ghostly fingers – the horizontal displacement – which is known as *disparity*. The closer your finger gets to your face, the further they appear apart (the greater the disparity), and so your brain can work out how far away your finger is because it has to cross your eyes further to focus on it.

This idea was hacked long ago in 3D movies, where you are in effect watching two different movies at the same time – which is why it can make you feel sick if the alignment is off. Usually glasses are involved, initially with blue and red lenses (and the video playing was also blue and red – blue lens blocks out blue, red lens blocks out red), and later on with polarisation in the lenses. What they did was allow your left eye only to see the video meant for your left eye and your right eye only to see the video meant for your right eye. If filmmakers wanted to make an object, like a biting dinosaur, come out of the screen, all they had to do was horizontally displace the dinosaur between the left and right images and your brain would fill in the rest. The further apart the two images, the closer your brain tells you the object is to you. What was found over time was if the object appeared too close to you as the viewer, you would naturally cross your eyes to converge them on the object, which would make it lose its effect. So limits were placed on how far the objects could appear to come out of the screen. It's an amazing illusion that demonstrates our understanding of how our eyes and brain construct our 3D reality.

But what if you have an eye closed, like to wear an eye patch because you're a pirate, or in fact only have one eye? Do you lose your depth perception? The answer is yes and no. You lose the parallax effect from two eyes simultaneously focusing on the object, but your brain is so clever it has built up many other methods of figuring out how far away objects are and of localising yourself. For example, we automatically interpret the perceived size of the object versus the size we know it to be to indicate its distance away from us – if we're seeing a very small skyscraper building, say, our brain knows it's probably a long distance away. Our brain can also pick up on the tiny changes in our eyes as we change focus between things close to us and things far away.

But there's another small trick we inherently know, usually without ever having thought about it. It's the thing owls do to work out the distance to their prey before they swoop. They bob their heads around. And we sometimes do this too, moving our head or even our entire body left and right to try to see around objects a bit and figure out their location in our environment. This happens when the relatively small separation between our two eyes is not large enough for us to tell the distance to an object. It's another form of parallax, known as *motion parallax*, where we build up different perspectives on objects by moving. Those that are closer to us move further across our field of view than do objects in the distance. Instead of *simultaneously* seeing it from two different perspectives, you're achieving multiple perspectives over *time*. Yup, just like an owl! Give it a try, you owl-person you.

Our two slightly different perspectives from our eyes, or movement of the head when only one eye is taking in light, also help us to overcome blind spots. These are very small areas in our visual field that are obscured, actually from within our own eyes (I'm not talking here about an overtaking car not being visible to your side mirror). This is where the optic nerve fibres pass through the retina in one spot at the back of the eye, creating a very small area in the visual field of each of our eyes that is obscured. To test this out, use the letters R and L below.

R   **L**

First close or cover your left eye and look at this letter 'R' with your right eye. Depending on the distance your face is from the page you should still be able to see the 'L' – blurry of course but still there. Keep your focus on that 'R' and move the page closer or further away until the letter 'L' completely disappears from your vision (which should occur when your face is about three times as far away from the page as the letters 'R' and 'L' are from each other). Likewise, you can close your right eye and look at the letter 'L' with your left eye, then move the page until the 'R' disappears. If, however, you open both eyes when in either of these positions, the obscured letter reappears, as it becomes visible to the eye that was closed. This means we don't often perceive that our blind spots are even there. Much of our vision and visual perception often relies upon this combined vision from both our eyes being processed in unison by our brain.

Let's return to the lab in 2007 and my work on SAM. For nearly a year I've been working on *stereoscopic* cameras for 3D vision for SAM. I'm using a pair of cameras inspired by our very own eyes, placed next to each other with a short distance between them. I program the system to analyse their data using the *sum of absolute differences* correlation algorithm. The two simultaneous 2D images can be used to calculate the disparity of individual objects between them and thus the distance a given object is away from them. The result is a low-resolution

3D point cloud, indicating where all the pixels the cameras have collected sit in 3D space – think of it as like a dark room you can move around in, with each point in the cloud represented by suspended, illuminated, coloured sand.

Later that year, working on my mind-controlled wheel-chair, TIM, I couple stereoscopic vision with *spherical* cameras to create a system that can see everything all around in 360 degrees. This is unheard of at this point in time. I have to figure out an effective method of combining the data from these two different camera systems to give the smart wheelchair a great way of observing and navigating unknown environments, as well as avoiding unpredictable objects and people along the way. As for many great engineering designs, I look to systems in nature for inspiration. I initially think the idea of eagle vision would be awesome, but quickly realise a bird of prey might not be the way to go. Predators tend to have vision honed for the attack, but what I am looking for is something defensive, so that my wheelchair avoids collisions and creates safe pathways of travel for the operator. Looking through a range of veterinary journals for information on various vision systems, I finally stumble upon the perfect biological inspiration – horses!

As it turns out, equine vision is very effective. With elongated pupils, placed towards the sides of their head, they can see in 3D in front due to the binocular overlap (as we do), but can also see much of the environment around them too. While their heads are lowered, they watch what's going on around them, which is vital given they can spend up to 70 per cent of their time grazing. Although the vision around the sides is not 3D, they only need it to detect predators, so they can leg it right outta there if and when they do.

Human eyes have about 120 degrees of binocular field of view (FOV) – the width of view we can see with both left and right eyes at the same time, and about 190 degrees of total FOV vision. When we look straight ahead and stretch our arms out to both sides we should only just be able to see both sets of fingers at the same time if we wiggle them. Interestingly though, as we're seeing them with the edges of our vision, they'll only be in grey-scale, as our retinas can't process colour from our peripherals.

Horses have their eyes towards the sides of their heads so they have a smaller binocular FOV than us – 65 degrees – but a much greater total FOV of close to 350 degrees, which is not far off the full 360 degrees. This means that when horses are looking forward they can pretty much see everything except their own arse! Told you it was an effective vision system.

So I model the wheelchair on this system while removing the blind areas behind, resulting in a 60 degree binocular FOV and a 360 degree total FOV. And so an equine-inspired vision system becomes the eyes for my smart wheelchair.

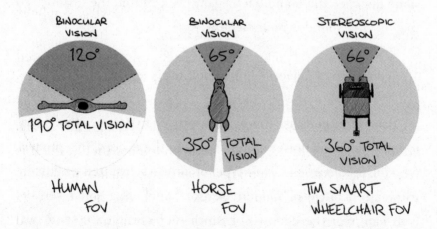

Figure 6: Vision field of view for human, horse and TIM smart wheelchair

Another commonly used concept from nature, brought into many of today's devices, is the time-of-flight principle. Humans, bats and dolphins all utilise this to estimate various distances. Think about those moments when you've yelled in a large empty space or even on a cliff. The greater the distance the sound has to travel before it bounces off objects and returns, the longer it takes to hear that echo of your own voice yelling 'COO-EE' or whatever call you make. This is the time-of-flight principle in play. If you're on a cliff, you yell the sound out, it travels all the way to the nearest cliff wall facing back at you, hits it, bounces back, then travels all the way back to you. Because of this distance in travel, you might wait a few seconds to hear the echo.

That time to the echo can be easily used to determine the distance, as we already know the speed of sound, which is a constant in air of about 343 metres per second, or 1235 kilometres per hour. If we multiply the speed of sound by the time taken, and then halve it because the sound has travelled the same distance there and back again, we can find the distance to the cliff the sound is bouncing off:

$$\text{distance to object} = \tfrac{1}{2} \times \text{echo time} \times \text{speed of sound}$$

Bats sometimes use this exact approach to perceive their environment, making noises and listening to the echoes, in a process we call *echolocation*. This type of process inspired engineers to create a range of modern sensors, including sonar sensors (you can hear the clicks – or pings – of sound they send out) and ultrasonic sensors (the pings are outside our hearing range), which can help robotic devices figure out where objects around them are located. These are also used on the back of many of today's vehicles as the reversing sensors. Incredibly, some humans have even developed abilities to use echolocation to, in effect, see through sound. I met one such person in my first ABC documentary, *Becoming Superhuman*.

Chris has been completely blind since birth. From a young age he developed an ability to echolocate. This is largely thanks to how our brains can rewire themselves, a wonder known as *neuroplasticity*. This phenomenon is one of the most important breakthroughs in modern science for understanding our brains, showing that they can change continuously throughout our life, though the plasticity is exhibited to a greater degree in children than in adults. These changes are physiological, occurring as a result of the brain adapting to interactions, learning, environment, thoughts and emotions, even damage and trauma.

The way Chris echolocates is an amazing display of neuroplasticity. How he perceives the objects around him is closer to the echolocation of bats and more advanced than most ultrasonic or sonar sensors today. It even gives us clues as to how the technology can be improved upon for future uses. Chris makes sharp clicks with his mouth and listens to the echoes with a seemingly superhuman level of sensitivity in hearing. He can not only figure out how far away objects are from the echo, which is everything I was hoping to witness upon meeting him, but he can also infer information about what he's bouncing sound off, which I found completely astonishing.

Based on the quality of the echo, Chris tells me everything sounds different – concrete, plants, trees, humans – so he has a very good idea of *where* the objects are placed in his environment, as well as *what* those objects are. When the objects are too far away for his mouth-clicks, he uses a cane, giving it a quick pound on the ground to make a louder crack. We were standing a whole 25 metres away from a building he was facing when he demonstrated this. He cracked the cane on the ground and listened, and then repeated and listened. His next words were simply awesome: 'Okay, it's 25 metres to the wall ahead, but if you look up it's 35 metres to where the building comes out from the wall,' even telling me that the windows and shades up there sound really cool. He was spot on. I couldn't help but laugh in amazement.

This same time-of-flight principle is the basis for lidar, which is much, much faster than sonar and ultrasonic, because it bounces light rather than sound, and the speed of light is nearly 300 million metres per second – about 875,000 times the speed of sound. It does, however, have trouble seeing windows, as the light tends to pass right through. It's also expensive compared to ultrasonic sensors. But every sensor comes with its own set of advantages and disadvantages. No sensor is perfect. This is why many robots today utilise sensor *fusion*. They'll be designed with many different types of sensors – such as lidar *and* ultrasonic – that can work together to tackle many scenarios.

Driverless vehicles are a good example of this. They are built with many different types of sensors because there are just so many different tasks they need to complete and situations they need to adapt to. A driverless car might have a range of sensors, incorporating cameras with AI that can see in colour and recognise such objects as traffic lights, people, lane markings and street signs; radar for the fast and accurate detection of other cars on the road around it; lidar for fast 360 degree mapping of the surrounding environment around, including gutters, trees and other solid obstacles; and ultrasonic sensors (as most new cars already have for reversing) to detect lower obstacles closer to the car that the other main sensors have no line of sight on.

Be it a robot, sensor, algorithm, system configuration or mechanical operation, many of our technological advancements have been inspired by the wonders of nature. Maybe our next advancements might even contribute to restoring our natural world. So the next time you're looking for boundary-pushing inspiration, have a look at what nature has already invented for us. It has had much more time to perfect it than you or I can even comprehend.

With millions if not billions of years
of creating, changing, updating, evolving,
nature inspires our most sophisticated inventions.
So let's guide those inventions to return the favour,
and use them to repair our natural world.

12

IS TECHNOLOGICAL TELEKINESIS REALLY POSSIBLE?

I WANTED TO BUILD A MIND-CONTROLLED WHEELCHAIR

'AHHH! IT WORKS!' I yell out in excitement, but no words can actually describe how I'm feeling in this moment. It's February 2008 and I've just connected *mind control* over my smart wheel-chair, TIM, with a custom brain–computer interface (BCI). I've associated three different thoughts with three different actions and used the distinct brainwave changes when you open and close your eyes as a method of confirming desired actions. These thoughts are all mental tasks synonymous with IQ tests, like arithmetic, writing and figure rotation. The parietal lobe of the brain works hard on these types of tasks, so I have some of my BCI electrodes connected to this area on the scalp, as well as to the occipital lobe at the back of the head, which is responsible for visual processing of all the data our eyes send to it.

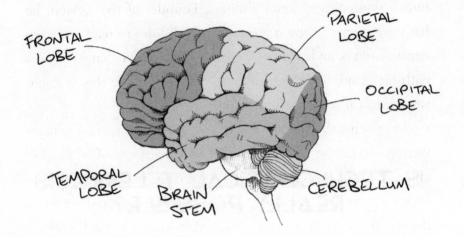

Figure 7: Lateral view of the brain

I'm wearing a headband fitted with a stack of electrodes picking up the EEG electrical activity of my brain from the surface of my scalp. Although it isn't amazingly accurate since the signals are not being taken directly from the brain, it's enough for my AI program to recognise the difference between these three thoughts, a heavy blink and when I'm thinking anything else (bucketed into an *other* class that leads to no response or action being taken). I've just imagined the thought that I've associated with *forward* for the wheelchair, the small screen connected to TIM has displayed that I intend to travel 'Forward', and . . . the wheelchair moves forward! I quickly get out of the chair, as the signals are transmitted to the computer on the wheelchair via Bluetooth so I don't need to be sitting in it, and I continue to make the wheelchair move around the office space. My mind is blown – this is some real-life Professor X stuff right here!

Professor Charles Xavier (Professor X) – a fictional character from the Marvel *X-Men* comics – is a mutant, a subspecies of human with superhuman abilities. Founder of the X-Men, he has exceptional telepath powers, with abilities to read and even control minds and matter. He has a wheelchair that he moves with his mind, and this is exactly what it felt like this moment was. The stuff of superhumans.

In creating the BCI electronics for TIM, I found that collecting *brainwaves* isn't too dissimilar from the design of an electric guitar. When you strum the strings on an electric guitar you create little vibrations in the strings and the electrical pick-ups collect these signals through electromagnetic induction. Each pick-up has small magnets with conductive wire coiled around them, creating an electromagnetic field, and when the magnetised strings vibrate within the electromagnetic field, they induce small electric currents to flow through those wire pick-up coils. You can turn up the volume to amplify the signals. You can filter the signals and make them sound clean – through amp speakers that convert the signals back into sound through vibrations. And this is all similar to how a BCI can acquire usable signals.

Electrodes placed on the scalp pick up the brain's tiny electrical EEG signals – in the order of microvolts (millionths of a volt) – amplify the signals to make them more usable and filter the signals to clean them up. Instead of electromagnetic induction to supply the analogue electrical signals, we instead use a small electronic circuit that picks up the electrical activity of our own brain at the sites the electrodes are placed. The resulting signals are then transmitted into the computer for processing. My programs and inbuilt AI make sense of it all and recognise the patterns of thoughts they have been trained to look for.

A week after this initial success, an academic group visits the university from Japan and I'm all set to give a demonstration of this new form of wheelchair control. Everything is set up, I'm wearing my headband, I've checked my connections, I've tested the wheelchair ten times beforehand, and I'm feeling confident it should all work. Everyone is standing around watching as I sit in the wheelchair facing away from them so they can see my screen. I focus my mind on the thought for *forward*, instantly evoking in my mind's eye a large 3D Rubik's cube in front of me, rolling forward. The computer responds, recognising my EEG patterns as familiar and falling into the *forward* class, so the screen shows that 'Forward' is being selected. TIM gives an audible *click* as the wheel motors kick in. I take off smoothly, travelling forward along the office, with a long glass wall to my left. I make it to the other end and give a command-blink (heavier than an average blink) to stop.

I now focus on simple addition arithmetic. Imagining the numbers appearing on the right side of me. The 'Right' selection lights up on the screen and I rotate the wheelchair 180 degrees until I'm facing the group of onlookers at the other end of the office. A moment later I'm heading back in their direction, but before I get to them my slightly skewed travel direction means I've come very close to the wall of windows, now on my right. I command-blink to stop the wheelchair and unintentionally create a new simple lesson in this form of control interface. A sort of *Mind Control 101*.

I'm looking over my right arm at how close I am to the wall and my inner voice says *Don't think of the thought that makes you go to the right!* Well, if I tell you 'Don't think of a purple elephant', what does your mind do?

BAM!

I seem to be getting good at crashing my creations. Before I can physically react and shut off the power, TIM takes off to the right and straight into the window wall, bending the right footrest and armrest inwards, but luckily not breaking the thick glass wall.

I quickly start to realise that there's a fundamental difference between controlling a device directly from thoughts and through physical movement. See, if you're driving a car you can think about potential problems, you can even imagine turning the wheel in a certain direction and crashing, without actually doing it. When you're speaking with someone face to face, you can be saying all sorts of things in your mind, but you still (generally) filter these things before you actually speak. This is all thanks to our prefrontal cortex, which in essence is our *filter*. It provides us with a sense of the future consequences of our current activities, gives us inhibitions, helps us with expression of our individual personality, assists in decision-making and moderates our social behaviour, letting us know what may be socially acceptable and what we believe is not. This is also, interestingly, an area affected when we drink alcohol, possibly leading to singing, fighting, speaking without a filter, calling those you shouldn't call or feeling like it's a good idea to dance on tables.

Essentially, using your mind to control a device means that you're going from thought and imagination directly to action, bypassing the *filter* of the brain. I learn this from crashing my TIM wheelchair in front of this group of academics, but immediately start working on combining this form of control with vision and smart robotic capabilities that will allow the travel to be safe.

The resulting tech means TIM can be controlled by thought and will also automatically avoid potential accidents, so in effect I'll be adding an automated *filter*. But while it's one thing having TIM respond to *my* thoughts, the massive challenges come with attempting to get it to work with other people.

My first set of participant trials see two testers do very similar things – focus on one thought and get themselves stuck spinning in circles. Once you're spinning, it's not easy to focus on stopping or changing commands. I cancel the trials and go back to the drawing board, learning from my mistakes and recognising that if you're ever designing something for use by people, the design needs to be intuitive and easy to use (especially for something this complicated). When the day arrives to get test runs going again, a mate turns up to my office and I take him through all the features. Being a wheelchair user, he quickly becomes very excited.

'This is so cool,' he says. 'Let me know when I can come in and give it a test run.'

I happily respond, 'You can give it a go now!'

'No, I can't.'

'Oh, why not?'

'Because it doesn't have a headrest. I need a headrest to hold my head in place and upright.'

'Ah.' Well, now that it has been stated it's pretty obvious, but I haven't previously given this thought. An awkward moment of silence follows. We both look back at the TIM smart wheelchair.

'And a seatbelt. I need a seatbelt to hold me in the chair.'

'Ah.' Again, this is all I can verbalise while processing the things I've missed.

The revelation I find myself facing is that to truly innovate, you need to keep in mind who you're designing for, make a prototype of your wild ideas, then include the people you're designing it for in the journey once you've shown them a glimpse of the future.

I start making sure I have necessary additions on hand, and when it comes time for the first trial participant with a disability to test out TIM, I'm nervous I'll be faced with another false start. We work together transferring him from his own wheelchair into the TIM wheelchair. I carefully set up the electrodes, each of which requires application of a conductive gel and correct positioning on the scalp. I move his hair out of the way for each electrode location, ensuring a good connection and making sure to apply enough pressure to hold that connection between electrode and scalp. From the front it looks quite neat. At the back of his head my mate now looks like a cyborg, with various colour-coded wire loops with tags sticking out from the headband holding it all in place. We test out the quality of the connection with a couple of mental tasks and go through the operation and control interface being displayed on the screen in front of him. I introduce him to the obstacle course he is going to control TIM through, allowing him to see he has to make a mental command whenever there is a direction change. He then has to trust that TIM will avoid all collisions in those directions. It's time to begin!

He goes silent. Concentrating. Focusing on the mental tasks to command TIM. *Click* – the familiar sound the motors of TIM give off whenever they go from a stationary position to the start of motion. And he's off! Travelling forward yet staying calm and focused. He commands the wheelchair to turn right with good timing. The wheelchair takes the turn around the corner, avoids an immediate collision and then the next few obstacles in its path. Left, right, left and then it straightens out. With a small stop zone, my mate times it perfectly, commanding the wheelchair to stop, and the first short test is successfully completed. He looks at me and smiles. We both just sit for a minute, taking in what has just happened.

It's a powerful moment as I realise that through action and persistence, big ambitious ideas can be realised. We have just shared a unique, incredible experience, and our minds have been expanded along with our bubbles of possibility. I am starting to realise that this deep sense of fulfilment and new-found purpose is greater than anything I could ever make for myself. And I can feel I am being changed forever.

Over time, as TIM improves, I try a variety of different thoughts for control, eventually landing on designs where you can focus all your energy and imagine movement in a particular direction. With numerous options to pick from and selecting the control operation themselves, participants are able to gain a quick understanding of how TIM works.

In 2011 I conduct successful trials with twenty people. Each test is just so encouraging. My Russian mate Albert, who has a C6-C7 spinal cord injury from a motorbike accident, learns fast and becomes skilled at controlling TIM quicker than most. At this point in time he is a twenty-year-old civil engineering student. Only half a year earlier he saw me wander past one of his uni classes with TIM seemingly driving itself. Albert quickly moved out of the class and came over to say hi, curious about how it all worked. A few days after that, he was testing TIM

out around the university, giving me plenty of feedback as he learnt. I've built in upgrades over the intervening few months, and now Albert has a seamless level of control over this smart wheelchair. While in operation, it appears they have become one and Albert's control seems effortless. Originally only working for myself, this technology has been challenging enough to build, before learning to take the big leap to allow others to operate it too. All that hard work is completely worth it, and I instantly know I want to continue creating for the rest of my life.

Years of work on this robot wheelchair TIM meant 'he' had such a consistent presence in my life. He was there with me in my office, I'd take him with me whenever I'd go down to the food court for lunch or just at random times during the day. I was usually testing out his latest obstacle-avoidance capabilities or wireless method of control, but sometimes I was just getting him to stalk people. At least a few times a week I'd sit with my brilliantly vibrant friend Britta, who worked at the uni concierge desk. We'd watch the range of hilarious reactions from unsuspecting passers-by as they were followed around by a wheelchair with no one sitting in it. I may have matured, but I'll never grow up!

Later on, in 2013, about half a year after completing my PhD on TIM, I went back to visit. He was still in the Centre for Health Technologies, but a few post-doctoral students were working on a different design for him. They had pulled him apart and put him back together again, moving his cameras and equipment significantly. So much had changed in TIM's appearance and operation, he was no longer the TIM I knew.

I honestly felt so gutted. It was such a deep sense of loss this time, greater than anything I'd experienced with a non-human entity before. I didn't tell anyone because I didn't think they would understand. I mean, I was only just wrapping my head around it myself. As mentioned earlier, we can feel connection to non-living entities, and even though those feelings may not be reciprocated (TIM was not capable of such things), the emotions become real for us. Just like a child with their teddy. To me, TIM was a companion.

And it was through these emotions, along with a great range of adventures that felt like they were a merging of imagination and reality, that I'd learnt so much about biomedical technology and bionics, creating personal BCI connections between us and machines, spirited by strong human-centred purpose. A range of interface alternatives have emerged since, taking us into the realm of bionics, from invasive devices that require planting arrays of small electrodes on the brain itself and carrying the signals out through the skull, to thin sheets that can be semi-invasively planted under the skin on the scalp, to non-invasive types like TIM that collect the signals from the outside surface. These approaches to computer and machine control through BCIs and brain–machine interfaces (BMIs) will get more refined over time, as we move from the clunky, prototype-looking designs through to new, more techy, claw-like, non-invasive versions (such as the first widely successful consumer and research products from my friend Tan Le's company Emotiv – which I think look awesome), towards those you barely see at all and invasive devices implanted directly into the brain for those who opt for this approach.

The current method of collecting the signals from brain activity is a constant trade-off. The closer to the source you can get, the stronger and more accurate the signal, but also the greater the difficulty and risk involved. Other ambitious approaches have emerged over time as researchers advance the technology into new forms, like neural interfaces that are either implanted in the vascular system (i.e. blood vessels) of the brain to pick up electrical activity, or others meant to be directly implanted onto the brain itself. Invasive technologies obviously come with greater risks and challenges, but the signal quality of the data and amount of data that can stream out from these systems create unprecedented opportunities for understanding more about ourselves and unlocking many more secrets of the brain and body. They are also taking leaps towards theorised *human–AI symbiosis*. This is a strange theory of the brain and AI being so integrated that we become one and the same. Our individuality is what makes us who we are, so direct symbiosis is not something I'm yet in any way convinced of as having significant benefits. There are intense and controversial challenges ahead for these developments, that's for sure.

As the technology advances, we'll not only be able to collect signals from the brain, but also to feed signals into it. People already have a range of different ideas as to how this can work, which means we may need to start thinking about the privacy of our thoughts and how much we want our brains connected to the internet and global networks. We've seen enough data privacy breaches to know that having everything streamed out of your brain could be a wild mess. To eventually feed signals into it too brings up visions like the downloading of skills to the brain in *The Matrix*.

Audacious ideas like these have been around for years, and for a long time have sounded like something you'd only ever expect to see in the movies. Things like brain implants collecting our thoughts, taking control of our actions, feeding knowledge and skills directly into the brain, or even uploading our consciousness to computers so we can preserve personalities and immortalise ourselves in a virtual realm. Whatever the concept and no matter how far-fetched or sci-fi it may seem, we as a species are inching closer to fusing with our creations.

13

HACKING OUR SENSES

SEEING IS NO LONGER BELIEVING

'This . . . this isn't real?' Neo asks Morpheus in the film *The Matrix*, as they stand facing each other. Between them sit two apparently age-worn couches, a small table and an old-fashioned tube TV. There is nothing else anywhere around, above or below, but whiteness. Complete whiteness. This is the first time Neo finds himself, knowingly, in a simulation. He asks this question as he runs his hands along the back of one of the couches, clearly registering believable sight, touch and texture. Morpheus replies in a manner that displays how well he knows the truth. 'What is real? How do you define real? If you're talking about what you can feel, what you can smell, what you can taste and see, then real is simply electrical signals interpreted by your brain.'

Doesn't that send your mind into a spin? Our individual reality is an accumulation of the things we've experienced and of the things we sense. Some similarities we have with technology lie in the fact that just as robotic sensory devices collect real-life information about their environment, encode the signals into streams of ones and zeros, and send them to a computing system to process (as we saw in Chapter 4), our own bodies do something extremely similar. We take in information about the real world through our five major senses, which convert the information into streams of electrochemical signals to send to the brain, and it is there that the signals are processed into sense and meaning. Earlier I talked about using these systems as models for our own tech, but just how well are we able to control our senses and, in effect, our reality? How can we hack our senses and trick our mind?

This comes back to biomedical technologies and how sensory signals can be fed to our brain. I've decided to categorise the possibilities into four technologically facilitated levels. They're levels of progression just like a computer game simulation, because hey, we can't prove we're not living in one.

A HUMAN'S GUIDE TO FOUR LEVELS OF SENSORY HACKING (SH-4)

Level 1: Sensory stimulation

Many current technologies that simply stimulate our senses, e.g. TV, music, haptic vibrations

Level 2: Sensory bypass

Advanced technology that bypasses the mechanics of our senses and sends similar signals through the various nerves to the brain, e.g. cochlear implants, bionic eyes

Level 3: Sensory injection

Technologies we haven't yet achieved that could plug sensory information directly into the brain, e.g. virtual experiences the mind interprets as reality

Level 4: Posthuman sensing

If our existence transcends our brain but we can still sense. We've well and truly crossed over into sci-fi territory by this point.

Our perception of reality, our individual experience of reality, is a funny thing. Take our vision, for example. It's limited. Electromagnetic radiation is a spectrum of waves that carry electromagnetic radiant energy at different frequencies, and propagate through space as radio waves (at the low-frequency end of the spectrum), microwaves, infrared (IR), light (the parts of the spectrum we can see), ultraviolet (UV), X-rays and gamma rays (at the high-frequency end). The visible spectrum of light that our eyes can see only accounts for a small slice of the various electromagnetic frequencies, with IR and UV light just outside our visual range.

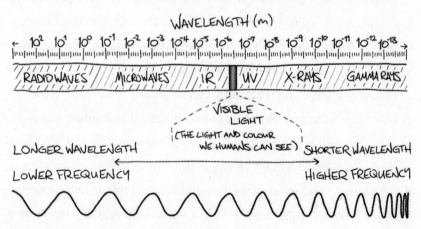

Figure 8: Electromagnetic radiation and frequency

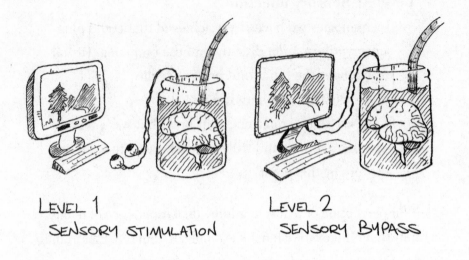

LEVEL 1
SENSORY STIMULATION

LEVEL 2
SENSORY BYPASS

Figure 9: Levels of sensory hacking (SH-4)

When light comes into our eyes, they convert it into a stream of electrochemical signals that are sent along the visual nerves to the brain. The brain processes these signals and constructs them into a form that gives us our sense of sight. Our brain is, in effect, locked in a dark jar that we cannot see, taste or touch. It is continuously fed all sensory information by electrochemical messengers that encode the information coming in through our senses, then are run to our brain for it to interpret.

In the case of our eyes and vision: Level 1 sensory stimulation includes computer monitors stimulating our sense of vision as our eyes observe the screen; Level 2 sensory bypass involves technology like a bionic eye bypassing the natural sensor that is the eye and sending signals along the optic nerve to the brain; Level 3 sensory injection is further into the future where the technology sits inside the skull (likely implanted in the brain) to

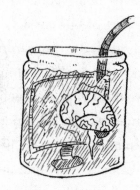

LEVEL 3
SENSORY INJECTION

LEVEL 4
POSTHUMAN SENSING

feed vision, injected straight into our perception of reality; and Level 4 posthuman sensing encapsulates how we might see in a philosophical future if we have transcended our biological body, perhaps existing in a virtual world and having transferred our consciousness into a computer.

Let's explore each of these levels in more detail.

LEVEL 1: SENSORY STIMULATION

The stimulation of hearing is quite easy. We seem to have had an amazing grasp of it throughout history, with and without the use of technology. When we sing, play instruments, play music through speakers or headphones, we stimulate our sense of hearing.

Our hearing is an intricate little system for transferring soundwaves into electrochemical signals that can be sent to the brain. It starts with the soundwaves being concentrated and collected by our ears (the visible part of our ear acts like a satellite dish). They then move through to the eardrum, causing it to vibrate. The movements are translated through small bones in the middle ear that move through to the cochlea. This is like a snail shell deep in the inner ear, full of fluid and tiny, delicate hair cells. These hairs run along the inside of the cochlea and are stimulated by soundwaves to sway and move. Low frequencies are slower, deeper sounds like what you'd expect from a massive demon of the underworld, and high frequencies are faster, high-pitched sounds like you'd expect to hear from a fairy. These different vibrations flow through the interstitial fluid as movement inside of the cochlea and are absorbed by the fine hairs within, each responding to different frequencies.

When those fine hairs are stimulated, they create electrical impulses that are shot up the auditory nerve to the brain. So when we listen to music through headphones, this is the culmination of many, many generations of humans realising they can hack the auditory part of the brain in a pleasurable way, by stimulating our sense of hearing.

LEVEL 2: SENSORY BYPASS

Our developments at this level over the past century have mostly come in the form of bionic ears to bypass the mechanics of our hearing when damaged, bionic eye research to do the same for our vision and nerve feedback from bionic limbs. These designs tend to bypass the sensor itself, instead sending their own signals through the relevant nerves to the brain. To possess such technology is to be partially cyborg. Again, we'll take hearing as a great example here, because it has had the most success in sensory bypass to date. Unlike a hearing aid, which firmly sits at Level 1, as it basically amplifies sound to make it louder, the cochlear implant is an auditory-bypass system and so has earnt its place here in Level 2. The genius of this design is to completely bypass the mechanics of how we hear for people who have more severe hearing impairment than a hearing aid can assist with.

The cochlear implant (also known as the bionic ear) takes a different, amazing, technological approach to the mechanics of hearing. The first component is a device containing a micro-phone that sits on the skin behind the ear. The microphone picks up sound and digitises it. This digital signal can be processed to bring out the most relevant types of sound, like the voices of people talking to you, while dulling out arbitrary background noise that might interfere. This sound data is then converted into radio waves and transmitted through a little circular disk, magnetically connected on the back of the wearer's head. The radio waves are received by the cochlear implant under the skin, converted into an electrical signal and sent to a snake of electrodes that has been pushed into the cochlea. The different frequencies of sound will cause their associated nerves to fire, sending electrical signals directly up the auditory system to the brain. This whole process has largely bypassed the mechanics of the ear and hearing system.

This level is synonymous with the beginnings of the trans-human era, where we become integrated more and more with our technological tools, our machines and our AI.

LEVEL 3: SENSORY INJECTION

Now, Level 3 is one we have not yet reached. It will require a significantly advanced understanding of our own brain and how it works. It will also require advanced technology to control the brain directly. The concept here is of plugging virtual worlds and streaming experiences directly into the mind – like the Matrix, minus the whole humans-becoming-slaves issue. It's an astronomical leap forward from Level 2 technology that stimulates the nerves carrying signals to our internal CPU, to Level 3 technology that overrides the sensory system, injecting the brain directly to hear what we want it to hear, see what we want it to see. This level represents the future we have not yet reached. It's well and truly down the pathway of the *transhuman era*: a philosophical time we are approaching in which biology and technology interweave and the human condition, intellect and physiology is transformed. This advanced Level 3 technology might sound scary and the stuff of futuristic fantasy, but that was once the case for bionic implants, flying vehicles, robotics, AI or even personal computing devices. Just because Level 3 hasn't happened yet doesn't mean it isn't possible or won't become reality within a generation. On the contrary, I believe we will see these concepts come to life. But do we really want it (even if it comes without the slavery)? If being able to live a virtual life of adventure beyond the body's capability is on offer, particularly in the future of palliative and end-of-life care – some people would choose the transformation with such science and technology to continue experiencing a different form of life.

This concept of injecting our senses seems like a moonshot development and one currently not quite within reach. So, are there possibilities even further beyond this? Well, we could go a quantum leap further to Level 4, which no longer even involves a brain. This is the point in the road where you've reached a large road sign that reads, 'Welcome to the Posthuman Era'.

LEVEL 4: POSTHUMAN SENSING

This philosophical era at some arbitrarily distant point in time and space is where our imagination can really start to run wild. In the posthuman era, humans will have transcended their traditional physical form and will likely no longer be 'human' by our current definition. There is much speculation concerning whether posthumans and humans could co-exist in this era, but the idea is that our advancements in fields such as nanotechnology, AI, biotechnology, gene sequencing and digitisation or quantisation of humans, could eventually lead to a merging of human and AI, or an uploading of consciousness. We may then live on as a new form of human – experience, thrive and consume in a completely different fashion, possibly sustainably and possibly in a manner that can ensure our survival, in some form, well into the future. This is a concept that of course brings up significant debate, from existential, philosophical, ethical and religious standpoints.

However fanciful this era might be, it has nevertheless been a dreamt-up future for our advancements at the intersection of humanity and technology. And that means that Level 4 must be noted in this list as what lies beyond Level 3, no matter how wild the concept. It is the level where the human can sense without a physical brain. In the case of consciousness-uploading, you may find yourself in an avatar form in a virtual fantasy world, looking around and listening, moving and touching, smelling and tasting.

Yet you will no longer own a physical, biological form in the real world. You will be completely virtual. Computing and storage systems upholding such technology would still need to exist and be maintained in the real world, powered by energy systems of the future, going beyond super-efficient solar cell technology and well into what I'd say are two holy grails of future energy: hydrogen-fusion energy generation coupled with ultimate forms of energy storage – think super batteries. Upkeep, security and maintenance of these systems would be a role for advanced AI (with robotic systems) to allow all humans the option of living on virtually.

We could keep our populations and reduce our burden on the earth – death as such will be drastically reduced, but we sacrifice reproduction in the process. So not all humans will take this path, or make the move later in life. Posthumans that have moved into virtual worlds will require Class I avatars, which I'll discuss in Part V of this book. Unlike in *The Matrix*, where we've had a virtual world plugged into our minds and pulled over our eyes unknowingly, here we instead accept these lively virtual worlds as another form of reality, and plug ourselves in willingly. We would require guides to help us navigate whichever of the seemingly infinite created worlds, holidays, explorations, challenges and adventures we would want to experience. There will always be endless things to do, and still plenty of purpose that we create for ourselves.

You will sense these worlds in a similar way to the biological systems of Level 1 or the technological systems of Level 2. Signals that represent the sense will be sent to your mind through your brain, only now your brain is also virtual. Your virtual senses get virtually collated and constructed, and your virtual consciousness experiences them, in a not too dissimilar way from how your current biological senses are biologically collated and constructed and your biological consciousness experiences them.

What could possibly be beyond this? It's like trying to think about what's beyond the edge of infinity. I imagine that we would have to exist as pure energy and still be able to sense our surroundings. Many believe this already happens after death. So that level crosses over to the other side. Beyond biology. Beyond technology. Into the realms of spirituality and religion. Realms I will not include here with the advancing technological levels of sensing.

What this framework of levels does tell us is that we are in the infancy of our available sensory hacks, since the majority of our technologies in this space currently fall into Level 1, and some senses are still lacking here as it is. Apart from hearing, Level 2 technologies are still in the very early stages.

What we do know is that our evolving technologies that continue to bridge the divide between human and machine are advancing and growing in number. You might think you wouldn't be interested in going down this path, but think about this. Would you accept replacing a damaged organ, sensor or limb with a technological counterpart? When these options become more readily available and successful, and you require a replacement to survive, the response probably becomes quite simple.

Everything we've ever known.
Everything we've ever experienced.
Everything we've seen, or heard, or smelled,
or tasted, or touched.
Our entire individual existence is a
perpetual construction of what we call 'reality'.
All orchestrated by the most complex
processor we know of:
the human brain.

14

IMMORTAL JELLYFISH AND THE BIONIC REVOLUTION

BIONICS WITHOUT METAL, GEARS OR ELECTRONICS

So far we've discussed bionics in terms of electronic and biomechatronic tech, but there is another approach to creating technology to be implanted in humans – using living organic structures we have created and grown ourselves, drawing on biology and technology in a different way from the traditional form of bionics. It reinforces the fact that we should continue to look for insights and inspiration in nature to create a wealth of solutions for scientific and technological translation.

Simply knowing the immortal jellyfish exists opens our minds to the possibility that a living creature can cheat death. Its actual name is *Turritopsis dohrnii*, and it's one of most incredible organisms to have ever existed on Earth, seeming to defy all known laws of nature. Discovered in the 1880s, what makes this little jellyfish so special is that it is biologically immortal, meaning although it can be killed, it won't die from old age.

It starts its life as larva, called planula, which then settles on the sea floor and grows into a colony of polyps. From these spawn a number of our jellyfish friends, which in just a few weeks grow into adults about 4.5 millimetres wide. If it is ever injured, starved, comes under great stress or just grows old, the immortal jellyfish adult pulls a kind of Doctor Who stunt and transforms itself to cheat death. It reverts back to a sexually immature stage, shrinking back down into a juvenile polyp in a process known as *transdifferentiation*. From these it can not only start its growth again, but also spawn clones with the same genetic make-up, each of which can continue to follow the same life cycle, basically rendering this jellyfish, well, *immortal*.

It's fascinating to know that a biological creature on this earth has evolved an ability to escape death. The human species, through our self-awareness and concepts of both life and death, has been *chasing immortality* throughout the ages. It's possible that nature holds keys to longer life, but modern medicine has already given us the ability to lead significantly longer lives than our predecessors, and now we're reaching ages where the brain struggles to keep up. With one person developing dementia in the world every three seconds, we should no longer look at just increasing the *length* of one's life, but also its *quality*. Furthermore, we need to consider the environmental impacts on the Earth when we all live longer. Solutions can be found to all of these challenges if we keep searching, so while we may on the one hand uncover ways to increase length and quality of human life, on the other we must also work towards sustainable balance between us and our living home planet so that we don't drive all species, including our own, into extinction.

Stem cell research is our own endeavour into the adaptability of cells. Stem cells themselves start out in infancy and, through the process known as *differentiation*, can transform themselves into one of a range of different cells, depending on what's required. But can an adult cell once transformed be made to revert back into a stem cell, jellyfish style? Well, the 2012 Nobel Prize in physiology was shared by Shinya Yamanaka and Sir John Gurdon, 'for the discovery that mature cells can be reprogrammed to become pluripotent' – able to turn back into a stem cell that can again differentiate into one of a wide range of adult cells. These induced pluripotent stem cells (iPSCs) hold great promise in the field of regenerative medicine, as they can then become any other cell type in the body (e.g. blood cells, heart cells, pancreatic cells, liver cells, neurons) to replace lost or damaged cells.

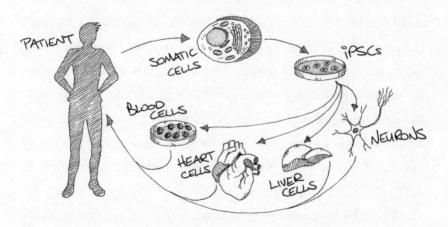

Figure 10: Somatic cells are taken from a patient, reprogrammed into iPSCs and then used to create other cells in the body.

More recently, stem cell research has found its way into assisting children born with physical disability to gain increased control over their limbs, regenerative medicine for people with damaged organs and attempts to repair spinal cord injuries by inserting stem cells into the spine.

Although this groundbreaking research is more in the realms of chemistry, biology and medicine than technology and engineering, merges of this science with technology are allowing us to create new forms of bionic systems and new solutions to repairing and enhancing humans. One of those methods involves additive manufacturing in the form of 3D printing. The standard version of this process creates physical objects by moving a print head and depositing a material (such as plastic polymers) layer upon layer, printing each successive layer to build up the 3D object. Now we have 3D *bio*printers, which use a 'bioink' that can be made of natural or synthetic biomaterials, alone or combined, and can include the likes of stem cells for bioprinting tissues. I myself have witnessed 3D bioprinting of liver tissue, deposited cell layer by cell layer. There has also been significant research globally into building heart, lung and muscle tissue and even nerve and spinal cord tissue. The field holds massive promise for organ engineering and regenerative medicine, and human-induced pluripotent stem cells represent a potentially near-unlimited cell source. The new form of cyborg is not what we envisaged in the past. Now the bionic parts are biomedical as well as biomechatronic.

When biology meets technology – this becomes a defining characteristic of the new era we're living in. We are improving on what we have while creating new ideas from scratch. Bionic limbs have been around in one form or another for decades, yet more recently, the research has taken new steps forward to give nerve feedback to some people who have prosthetics. One way this can operate is fitting a motorised prosthetic limb with sensors for temperature, touch and texture, whose data can be encoded and sent into a stream of artificial nerves up the prosthetic and into a special translator. This translator is a patch of nerve stimulation electrodes that has been inserted in the edge of the chest and connected with the nerve endings that would otherwise have sent signals to the brain from the limb. Through stimulating these sensors and finding out what the individual feels in response – touch on a particular finger, heat, pain and so on – the nerve endings can be mapped to various types of feeling and response. Once these have all been matched up, an individual with a prosthetic limb has the option of *feeling* what the limb is feeling, integrating human and machine that little bit more. It's another step towards the slow but steady spread of a global bionic revolution, a new era where our biology connects with technology, where technology becomes part of us.

But I believe the major turning points in this great integration between technology and humans will show themselves in a few key events. The first will be in the near future when a human chooses to lop off a functional limb to replace it with a bionic prosthesis. This may be due to the beauty that can be achieved in the form of it – not to say that our own limbs aren't a thing of beauty, but the artistic expression that can come with a prosthetic really will become appealing to many. (Just consider what we already do to ourselves in the name of fashion and vanity!) Alternatively, the main driver behind the replacement may be function, as the person in question may desire a limb that can perform different or more advanced functions from our own limbs.

A second turning point will be when a person wilfully exchanges a perfectly functional biological sensor for a technological counterpart. This could come in the form of hearing or sight, where the bionic sensor provides increased range or greater functionality in sensing than is typical for the sense we were born with. This second turning point is likely to happen well after the first, as human-made sensing technology still has a long way to go to adequately achieve Level 2 or Level 3 sensory hacking (see Chapter 13), whereas devices like prosthetic arms have already in some cases successfully reached functionality that a normal human arm cannot match. (Australian professional drummer Jason Barnes, for example, who has a motorised prosthetic arm wielding two drumsticks, has been labelled the world's fastest drummer.)

One thing is certain. We will see inspiring advancements that provide greater quality of life for humans in need, but we will also see shocking and surprising uses of these technologies as they advance. Either way, we're living in surreal times given the fact that humans are taking control of our own mechanics and even our own reality.

With our growing ability to replace what's under the hood, we've seen improvements to technologies such as bionic hearts, pacemakers, cochlear implants, bionic eyes, 3D printing of replacement limbs and even artificial pancreases and livers. With robotics and AI systems added in, we can also augment ourselves and add equipment to the outside of our bodies, with exoskeletons slowly becoming more useful, cheaper, more light-weight and smoother in mechanical movement. They provide us with superhuman strength, allowing an average person to easily lift objects many times heavier than they could normally handle. They allow us to stand and walk when our bodies cannot perform these functions for us. They will one day, in a few decades, make it possible for us to learn actions and movements by handing over control to exoskeletons – made using advanced materials science in addition to physical robotics – that we can wear like clothing.

We are creating new abilities to add to our human constraints and augment our limitations. We are creating new forms of replacements for what lies within the human body and facilitates its complex functions. We are taking control of our bodies, our minds, our lives and our mortality. Why? Because we are human, and we dare to dream big.

15

BECOMING SUPERHUMAN

HOW RILEY DROVE A CAR WITH HIS EYES

In 2015 I met a young man and now good mate, Riley, while I was working at a large charity that does amazing work in the disability community. I was told this young guy coming in to test out some communication devices was a bright spark with a great love for empowering technology and had contributed significant community work and fundraising himself, all before reaching high school. So I decided I'd go hang out and show him a few pieces of cool tech I was working on. Riley was born with high-level cerebral palsy, which affects his movements and his ability to speak. So Riley mostly communicates through his eyes, which are incredibly expressive, and also through some vocalised sounds that close family and friends can understand. In particular, the non-verbal communication between him and his father, Clint, is a wonderful thing to witness. To put it into perspective, one child is born with cerebral palsy in Australia roughly every fifteen hours and about 17 million people are affected globally. Often the parents have no idea what it even is until the diagnosis comes through. Cerebral palsy is a broad term and includes various movement disorders that appear in early childhood. Although individual causes are often unknown, it tends to be due to abnormal development or damage to the immature brain as it develops, most often before birth, but can also occur during childbirth or shortly after. Riley only has physical limitations. His mind is completely intact and he is a very bright and driven individual.

I took him through a few games I had been working on to be eye-controlled using a few off-the-shelf technologies, and then took him through his first virtual-reality experience using my own Oculus Rift DK2 headset. We had fun and some good laughs, and afterwards I was left with a sense that he was different from other people I've met, with and without disability. He was tuned in, motivated, driven. He had a rare level of determination and his ability to connect with people through his eyes was instant. I felt I could in the short time together already understand him non-verbally to some extent. This meeting stayed in my mind and fuelled my work.

Weeks on from meeting Riley, I find myself at my desk thinking about our interaction and wondering if technology could harness expression and movement of the eyes. I don't mean externally tracking the eyes with computers that may obscure line of sight. No, I'm wondering if we can harness the *power* of a person's eyes . . . of Riley's eyes.

It's a Friday evening a few days on, and I had planned on going out to the city with my then housemate Ben, one of my best friends from the age of ten. Instead, we decide to take it easy at home and I get to work building an early-stage prototype of my eye-tracking idea. I want to test the concept of using the electrical signals of the eyes through electrooculography (EOG). I dig up some electronics I have at home from my designs in thought-control (EEG) and muscle control (EMG) for my PhD and for other side projects I had been tinkering with since. I repurpose them for a simple version of eye control, starting with eyebrow movements. I connect one electrode above and one below my right eye. I then place a third electrode on the bony part of skin just behind my ear. This is a reference electrode that helps stabilise the signal I'm about to capture between the other two electrodes. I'm measuring the electrical difference between these two, which gets disrupted when my eyebrows move.

I record a few different words of me speaking and save the samples on my computer. A few hours later I have a program that can trigger these words through sequences of specific eyebrow movements. To make it visual, I program a circle on the screen to change and grow given fluctations in the signals from my movements. When the circle grows beyond set size thresholds a sequence of squares begins to appear and light up, with a new square in the sequence being added each second, up to six squares. The pattern of eyebrow movements then creates the sequence. Eyebrows down = empty square, eyebrows up = lit-up square. Sequences of squares facilitate the output, being various words. It operates in a Morse code sort of way. And before I know it, Eyebrow Talker is ready for a test.

First I'm going to try to communicate 'Yes' through my computer. Let's assume that an empty square is represented by a 0 and a lit-up square is a 1. The sequence I've made up for *Yes* is 001011. I raise my eyebrows, the circle grows and a sequence of squares on the screen starts lighting up. As I raise my eyebrows the squares light up and when I let them drop again empty squares appear. The sequence of six completes and . . . Success! My computer instantly says 'Yes' in my voice.

Ben turns to me. 'Did your computer just say that?' I look at him with a blank face, move my eyebrows up and down a few times like a weirdo and my computer blurts out, 'Yes'. We burst into laughter. With movements of my eyebrows I get the next voice samples working: 'No', 'Drink please', 'I'm hungry', 'Hello', 'Goodbye' and a few other random samples.

Now that this is all working and I have a prototype of how the electrical disturbances from the movement of eyebrows can control a computer, the idea can be expanded further. What I actually want to harness are the smaller electrical signals from the movement of the eyeballs. The EOG signals. Riley's EOG signals. If we can harvest the signals from his eyes to control a computer, we could control lights, we could control a fan or air conditioner, we could control a TV and much, much more.

I got to adapt this level of ambition further when I embarked on the adventure of my ABC Catalyst (and The Feds Productions) documentary with Riley and my small Psykinetic team (Nick, Louise, Maggie and Meg). In the show, Riley's mother, Casey, explains his cerebral palsy to us and tells us how, when he was young, he used to stare at the lights and try to make them turn on, and stare at the TV trying to make it change channel.

I say I know the feeling. You know it too, right? Those times you've stretched your hand out at something or stared intently, trying to see if *The Force* of telekinesis could be real. Casey says that one day while Riley was trying, the TV channel changed. He thought it was him, so he let everyone know he has a super-power. This leads to us naming the documentary *Becoming Superhuman*. In it, we attempt to create technology for Riley to achieve this power. I tell him we'll aim for him to be able to control lights and TV using the electrical signals of his eyes, but then I ask, 'If you could do anything with this technology that we're going to make, what would you do?'

He looks at the large computer in front of him that uses infrared and camera tracking to estimate where his eyes are looking, allowing him to type slowly using an on-screen keyboard. I'm not expecting him to already know exactly what he wants. I thought we might do something cool but safe, like remotely fly a drone. He selects the *speak* button and the computer says, 'I want to drive a car.'

'You want to drive a car?' I repeat as the voice in my head says, *Ohhh, shiii* . . .

'Yeah,' he replies. Although it takes effort for Riley, he can verbalise 'Yeah'. He sometimes makes a clicking noise as an alternative to *Yes* and can also verbalise 'No' well enough for most people to understand.

'You're only 13,' I say with a bit of a laugh, my mind starting to race as to how we're going to make this happen. He just smiles back. I ask him what he thinks about a toy car, remote-control (RC) style. He loves the idea. I remember how much I loved my first RC car too. Will it be enough? My mind answers this one pretty quickly: *No!* In this day and age we need to think big, to set the bar high. I let our producer and director, Julia and Wain, know I'm going to find a way. We're going to drive a car!

So we set out to do just that, while keeping secret from Riley that he might get to control something a bit better than a toy car. The best thing to do when you have a wild idea you want to take action on is to just start, then tackle each hurdle as it comes, figuring out the problem-solving along the way. If you try to learn everything you'll need and have everything planned out perfectly before ever taking the first step, you'll likely never begin. We rent a small warehouse where we can immediately get to work with all our equipment. I call my mate Darren in Perth, an engineer and founder of Dreamfit, an organisation that creates devices to help people with disabilities achieve new goals. I ask him about hacking a vehicle and he says he'll send me an off-road buggy from an unfinished project. Perfect! My team and I start drawing up a plan.

The idea starts with a unique headband fitted with a few electrodes to pick up the tiny electrical EOG signals of Riley's eye movements. The signals from his eye movements will be sent into a circuit similar to my brain–computer interface in that it will amplify and filter the signal to prepare it for the tablet PC. Once the digital stream is received via Bluetooth, it will move through an AI program trained to recognise what the signals look like when Riley moves his eyes in the different directions. My friend from university, Nghia, is helping us with the data capture and training of the AI. Based on specific eye movement sequences, like centre–right–centre, commands will be sent to an environment control unit (ECU) we build to control home equipment, and later repurpose for the buggy to send commands to control the motors. Riley would need to *adapt* with his *superpower*.

This is fast becoming an ambitious design, involving AI, biomedical technology, robotics, cloud services and computing, 3D printing (for the headband and other parts) and sheer human spirit.

In the lead-up, when I talk to Clint about our inspiration for some of these designs, he mentions that Riley finds it difficult to watch superhero or sci-fi films. He says Riley has high muscle tone in his neck and it sometimes involuntarily turns his head to the side from the level of sensory stimulation of such films. That means his eyes turn to the side too! What do we do now if this could also be the case when driving a buggy? Do we continue on with this idea, only to find he cannot keep his head, or eyes, facing the road ahead? Or do we cut this idea and go back to the drawing board?

As part of his preparation and for data collection of Riley's EOG signals, I've previously put dots on his home wall, one dot straight ahead of his eyeline for *centre*, and others to the *right* and *left*, *up* and *down*. Now I tell him, 'I believe you can gain control of these involuntary movements.' It's a tall order and I honestly have no idea whether he can, but I want him to try. 'Imagine you're doing something wild like driving a real car. Put yourself in that headspace, focus on keeping your head facing straight ahead, and practise moving your eyes between these different dots.' Off to work he goes. The next day Clint says, 'He got me to time him holding his head straight. He usually can't keep his head still for more than a few minutes but he lasted 45 minutes! Never seen anything like it!' Yeah he's got this.

Amazing what the power of our belief in someone combined with their own self-belief and determination can achieve.

Bit by bit, our various modules of the overall design are starting to work. There are still many hurdles ahead. The 3D printing of our custom headband is one of a number of things proving to be painful. Our software intern Nick tells me he'll stay back with me to work on everything we need to prepare, no matter how long it will take.

It's the night before Riley arrives to try the environment control. The internal workings of the headband are ready to go. They can pick up the EOG signals from the eyes, amplify and filter the signals then transmit them wirelessly to a tablet PC computer. Our AI program can figure out from there with a high accuracy which movements the user is making with their eyes, based on the pattern changes in the EOG signals. Our tablet PC will soon be made to *talk* to (send signals to and receive signals back from) a mini-computer (known as a Raspberry Pi). We've made this credit-card–size computer our ECU, which can turn on and off the lights, change the TV and control a range of other home devices. We're getting closer!

After a few hours of rest I turn up to the warehouse the next morning to find the 3D printer has finally completed the left and right sides of Maggie's headband design that will hold the electrodes and LEDs on Riley's head. We set up a space to look like his lounge room at home – a couch, TV, speakers, hanging lamps, two fans – and add some fairy lights for fun. We ready the communication between the tablet PC and the ECU, aaand . . . it doesn't work. Eventually, we realise one of our wires might not be providing power to the ECU. I rewire the little circuit board.

Success! A fan turns on. We quickly test a few more devices. All seem to be working. Okay, ready to go!

Riley and his family arrive and I start setting the headband up with the electrodes that stick to Riley's temples to pick up his EOG signals. It seems we have a problem. The headband has been printed to curl around the back of his ears but we didn't account for how tightly his head pushes against the headrest of his wheelchair. Plan B – we remove the headband for now and place the loose electrodes on his temples.

I talk him through his new technological telekinetic superpower. The various movements he can make with his eyes to command the environment include sequences such as looking straight ahead centre, then left and back to centre to toggle the selection between the devices he will control. Centre–right–centre turns the selected device on and doing it again turns it off. We kept it reasonably simple for this one and included up and down movements for adjusting volume and for changing the channel on the TV. Riley sits for a bit, concentrating. He looks centre to left to centre a few times and toggles selection to the hanging lamp. With an intense level of focus and determination, he makes a flick of his eyes to the right and back again, briefly closing his eyes once he brings them back to centre. The movement is registered by the system and as his eyes open the light magically turns on too. It works! Next, he turns on the fairy lights, then the left fan, then the right fan and then the one he really, really wanted . . . the TV. I cannot begin to describe this feeling and I imagine Riley feels the same way. After the long afternoon we celebrate our first success and see everyone out of the warehouse.

For Nick and me it's back to work. First thing is this headband, which needs to be reoriented to go up and over Riley's head. Armed with the not-so-high-tech solution of a small handsaw, and with only one real chance at this (spare doing battle with the 3D printer again) I remove the back half of both headband sides.

A new challenge has popped into my mind. If Riley is using his eyes to steer, it might make looking at his surroundings and assessing oncoming collisions a tad difficult. So using a bit of inspiration from observing Chris, the echolocator who has learnt to see through sound, we decide to use a similar form of sensory exchange. We're going to get Riley to see through touch as well, through his skin.

Our brains are incredibly adaptable, and this is what we hope to see in action. Riley will adjust himself into *sensing* his environment and *controlling* the buggy. The buggy will have a lidar rangefinder on the front, similar to how our UTS guide robot SANDRA perceived her environment, to quickly scan the surroundings and monitor potential collisions while the buggy is in motion. This information will then be sent via a cloud service to a little computer beside Riley that translates the data into haptic vibrations similar to those from a smartphone. It feeds these vibrations to straps on Riley's arms, two on each side, allowing him to *feel* the environment as he travels through it. Vibrations on the lower arm represent objects nearby, lower left arm for oncoming objects close to the left, lower right arm for oncoming objects close to the right. Oncoming potential collisions on the left side of the buggy vibrate the left upper arm, and on the right side the right upper arm. For all of these, the closer the objects, the stronger the vibrations.

Figure 11: How we put together Riley's buggy

This harnesses the power of the mind to adapt to different types of input, drawing on a small part of our neuroplasticity. So in effect, I'm designing this the way I would a robot like SANDRA, only here the main processor at the centre of it all is Riley's brain. To test it out, Nick and I program a small buggy simulator game with these vibration motors. I visit Riley to show him how it works and to give him the game, allowing him to test out the functionality of eye control while sensing through vibrations the environment ahead of the vehicle. He picks it up quickly and gets practising.

When the awesome off-road buggy arrives at our warehouse, the team and I immediately get to work on the integration process, taking the buggy down to a soccer field we've hired nearby. We cannot seem to get a line of communication between the tablet PC and the onboard buggy computer that controls the motors for the rear-drive wheels and the steering wheel shaft control. We program a different approach. Nothing. We reset the systems and connections. Still, nothing. Riley will be here in the morning to drive it, so we keep pushing on until we figure it out. On the plus side, we've unintentionally made the rejigged headband look much better than the original design.

At 1.20 am the new program is ready. Nick hits a key on his tablet PC that's meant to send a command via USB to the onboard buggy computer telling it to turn the steering wheel . . . and immediately the steering wheel rotates to the right! This means we can progress with getting all these systems communicating and creating a technological bridge between Riley's eyes and the vehicle motors.

Darren arrives to help and we keep testing. At 3 am we take the buggy back to the soccer field, set up and hop in. I get in the ready position. Then I send a few commands through eye movements. Centre–up–centre . . . and we're off. The vehicle is driving forward and I can make left and right turns too. It seems to work well but with very little light on the field, we're not sure exactly how well. At 4.30 am we call it a day.

We're back in the park at 6 am, setting up an obstacle course with large cardboard walls and witches hats for Riley to drive through. When Riley and his family arrive, I finally reveal the surprise. I say with a laugh, 'You know how you said you wanted to drive a car?' as we roll the wheelchair around the tent wall to reveal the buggy parked within. Riley's face lights up and I continue, 'So it might not be a car, but I think it's a little bit better, what do you reckon?' With an ecstatic smile nearly stretching the width of his face, he lets out a few clicks to say yes.

I show Riley his headband. We've added three blue flashing LED lights either side, not because they do anything mechanical, but because Riley likes the idea of looking a little bit cyborg. He's absolutely delighted. We put the headband over Riley's head and its two electrodes to his temples, then place a third electrode around the back of his neck as a reference to stabilise the EOG signals. We strap the vibration motors to his arms and initialise the computer systems and programs. The tablet PC mounted to the front of the buggy displays his EOG readings. We're ready to go.

First I direct him to focus and look ahead. 'Take your time. Left.' The team, family and friends all witness Riley's technological power as he flicks his eyes from centre to the left and back to centre. The steering wheel instantly rotates and the front wheels turn towards the left. 'Oh! There we go!' I yell out, relieved to see initial signs it's all working. 'All right, everyone, we're going to have a go.'

I'm riding with Riley for safety – I can take the wheel any time and hit the brakes if I need. I also have a big red kill-switch button next to me that I can press to shut the engine down if anything goes wrong. '3 . . . 2 . . . 1 . . .' and Riley times it perfectly with a centre–up–centre eye movement command. 'Wooooo!' I cheer as we take off, faster than I was expecting. The first turn coming up ahead is to the right but Riley is experiencing a surreal moment of sensory overload, like most people feel the first time they drive a car or ride a bike. He looks at the boxes ahead to his left that he doesn't want to hit and the buggy registers a left command. I hit the brakes as we roll straight into them.

As the day wears on, Riley keeps trying really hard. And he tries again. And again. And again. And he just keeps at it. But something is going wrong with our technology.

I start to realise that the loud motors that vibrate the entire buggy are also shaking us, which create disturbances in the EOG signal (called noise), making it jump around a lot. My mistake here was that our AI programs learnt on data from when Riley had been sitting in a still position, not when he was in a moving, juddering vehicle. Not only that, the front brake pads on the right side came loose and in the last run sheared themselves clean off. While Darren and the team fix the brakes, Nick, Nghia and I have a go at retraining the programs, incorporating the new EOG noisy data from when Riley was actually driving the vehicle. With three hours out of action due to Darren also finding a fuel-injection problem with the buggy, we have enough time to get the AI program updated.

The day is nearly over. The buggy is fixed and the new AI program is loaded. When I initialise the system, another problem reveals itself. The tablet PC, which has been sitting in the sun, is overheating and shutting itself down. Nick places it in the drinks fridge to try to hurry up the cooling. And now, the local soccer teams start arriving for practice. Riley, like us all, is becoming tired. His will and determination, however, remain completely fired up.

We secure the cooled PC back on the front of the vehicle. We have one false start where I jump in too soon with the kill switch. We go for one last try at a clean run. Riley's eyes remain ahead with a steely focus on the track and we're off. His eyes are bulging with sheer determination unlike anything I've seen before. They dart back and forth in command as he manoeuvres the vehicle right, straight, left, left, straight, right, left, straight . . . I'm trying to remain calm but can barely contain myself, letting out strange squeals of excitement. The final turn approaches. I can feel his focus is not waning whatsoever, but

I hold my breath. I feel the buggy rotate to the right as he times it perfectly. As we cross the finish line I throw my hand up and shout out, 'Yeeaah, we did it!' Realising we're headed for a fence, I quickly tell him, 'Up, up, up.' He gets this and moves his eyes centre–up–centre to slow the buggy to a stop. I'm laughing and cheering as he looks up, crying out in joy. This moment is instantly, and forever, etched into our memories. Riley has just made the impossible possible.

What an unbelievable adventure. It also showed us a glimpse of the implications for our future. We may continue to merge our biology with technology, in much more seamless ways than this. With human-centred purpose coupled with resilient will and determination, supported by advancing bionic and AI technology as driving forces behind the systems we merge with, we might all one day . . . become superhuman.

PART IV
ENVISAGE EXTENDED REALITIES

Soon enough one of them
may be your actual reality

16

LIKE PARALLEL DIMENSIONS

VIRTUALITY AND REALITY
ARE INTERTWINING

With an audience of 2000 sitting in front of me, I'm standing in the middle of a stage at a creative conference, slowly extending my hand towards the smart stage lights. I'm hoping I can take control of them to demonstrate the potential interlacing of virtual and real worlds as we move into the future. The lead-up to this moment has involved a lot of planning and a bit of imagination. The conference organisers sent me the stage layout ahead of time to build a virtual version of this stage on my computer at home. My audience in this virtual parallel world *are all large Lego people, the virtual stage and lights are all modelled on their real counterparts, and I have a cartoon-like avatar of me standing on this virtual stage.*

(See what I did there with the grey italic text when talking about experiences in the virtual world? This technique just hit me while writing this because, as my teams and I have found when we talk about it, moving between the two realms in conversation, and now in text, can be confusing. So, I'd like to introduce this technique as 'virtualitalics' – pronounced *vertu·ali·talics*. From now on these will appear whenever I talk about what's happening within the virtual world.)

A human motion-capture sensor is tracking my every movement and mapping them to my cartoon avatar, *so that he's making the same movements in the virtual world* that I'm making as I stand on this real-world stage. A large screen behind me provides a live feed from *a virtual camera on the edge of the virtual stage, angled to view my avatar, the lights and the Lego crowd.* The program has been coded so that *if my avatar points to the virtual lights, they'll change colour, until all lights have been changed, after which they'll move to face whichever direction my avatar is pointing.* It was then a process of learning how the real stage lights operate and how to control them. The virtual stage program continually sends out data to tell the computer controlling the stage lights what they should be doing, which is to mimic *the virtual stage lights* by matching their colour and movement.

This means that as I reach out my own hand to the lights near me, this should effect change in the virtual world by *moving my avatar, who has the power to change the lights in there,* which will in turn effect change by controlling these lights in the real world. I hold my hand out towards the four lights shining purple-coloured beams on my right, and for a moment . . . nothing. Is it stuck? I hold out my hand a little longer and move it around slightly. Behind me in the feed of the virtual stage, *the lights change colour,* and a split second later, their real-world counterparts change colour too! The purple beams become a cool *Tron* blue.

I breathe a sigh of relief and proceed to move my hand out towards the four lights on my left. These ones take even longer, and then . . . blue! They all change. Now that all eight stage lights are blue, they're supposed to follow the direction of pointing. Again, a wave of relief flows over me, hoping the next step will work too. I slowly move my hand up towards the ceiling to my right. Reminiscent of some sort of *Iron Man* missile targeting system, all eight lights move to the direction I'm aiming my hand. I pull my hand back and stretch it out up towards the left, and again all the stage lights immediately respond and turn towards the new position. I can instantly feel this strange connection between what's happening here on the stage and what's happening in the *virtual dimension* that has been interwoven with this live location.

You've probably already thought about the idea of alternate realities – parallel dimensions, like those in sci-fi and fantasy. If you're old enough, you might remember the 1990s show *Sliders* about travelling through wormholes between different parallel universes of the same world. It makes you wonder what 'you' in other dimensions would be like. In which ones are you kind and empathetic? Or completely selfish? How would your own life have changed given different opportunities and decisions? If you'd moved to another city, let that person who gave you butterflies in your stomach know how you felt, taken the plunge to quit your job and have a go at those things you always wanted to do . . . or, alternatively, if you *did* do these things, maybe you think about what life would have been like if you never had.

I know I think about this all the time. I used to have a fairly rigid mindset, sticking to the idea that *seeing is believing*. I felt I had a good grip on reality. Now I question my reality every day, knowing that things may be temporary, being open to challenging the things I think I know, and feeling like there could be something more, right in front of us, beyond what we perceive of the world. I don't think it would completely scare me to find out this is all just a simulation, or an experiment (depending on why and who's running it – sounds a bit *Truman Show*!) or one of many alternate realities (which actually sounds exciting). I'm not saying this is what I believe, just that I'm *open* to shifting my beliefs, questioning my existence, wrapping my head around advancements in science, and learning about the perspectives of others. But this isn't a section about the physics of a multiverse. No, this is about how we're now creating our own alternate realities, our own parallel dimensions, using nothing more than our imagination and today's technology.

Have you tried virtual reality (VR) yet? If not, try it! It may have appeared gimmicky at first, or if it was run on a system that couldn't keep up with the graphics (graphic-intensive realistic VR running on a smartphone, for example), it may have even made you feel a little ill and you haven't tried again since. If that's the case, don't give up. The systems have now caught up and the equipment is fast and powerful, providing smooth, mind-bending and truly immersive experiences.

From VR to augmented reality (AR) and some interpolated areas in between, the terminology surrounding these technologies is fast becoming hazy. All are part of the family of technologies known as extended reality (XR), which facilitate applications along the *reality–virtuality continuum*. In a nutshell, with VR you go into a virtual world by stimulating your senses (particularly vision) and almost forgetting about the place you're actually in. This can be incredibly immersive, helping your mind let go and feel as if it is within this digital environment. AR brings digital objects into a person's perception of the real world, sometimes hologram-like, viewed with the use of a devices like head-mounted displays or smartphones – think *Pokémon GO*. This is a bit closer towards the reality end of the continuum and falls under the mixed reality (MR) subsection of the continuum, but is rarely the case without adding some form of ability to interact with these holograms. For that, the combination of many types of sensors previously used in robotics, and even the addition of AI in certain applications, allows virtual objects and real-world objects to be attached to and affect one another. Users can experience many levels of interaction that blur the lines between the virtual and reality.

MR was what I was acknowledging on that stage, with the connected virtual world acting like a directly connected parallel dimension. We can now explore this continuum and we'll find ourselves spending different amounts of our individual time at different points along it. Some people will reject the technology

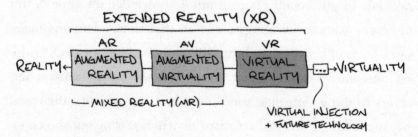

Figure 12: Reality–virtuality continuum

and live their whole lives at the real-world end. Some may reject reality and find themselves as far to the virtual-world end as possible. And many will dart in and out of the in-betweens as they see fit. This will become more common the more our jobs go beyond the digital space into the virtual space – tour guides and facilitators of experiences in virtual worlds; architects of virtual spaces, buildings and worlds; trainers of simulated education experiences; and designers, programmers, technologists, storytellers, creatives and regular workers choosing to do their work in these anything-and-everything-you-need environments. Roles we haven't yet even imagined will appear at various points along the continuum, in addition to its infinite range of applications.

Our imagination creates our own alternate realities. Think about those moments you've found yourself staring at the wall, out the window, at the sky or stars. Where are you in those moments? Where is your mind? Many of us felt that daydreaming in school meant we had difficulty learning when, on the contrary, our mind was longing to learn all the things – about life, about the unknown, about why the sky is blue, what holds the clouds up there, what stars are made of and why they only appear at night. Our imagination runs wild from a very young age, not only to pull together our limited knowledge on a topic, but to fill the gaps by creating stories so that these elements make some form of sense to us. I found I was always in my own world, my alternate realities. I'd let myself visualise cartoon-like characters running alongside the school bus and jumping off trees. My mind would drift to other worlds when staring out the classroom window. Daydreaming is critical to exercising our creativity.

Alternate realities have been made possible by understanding and manipulating this very concept for generations, stimulating our sensory inputs and learning to gain some level of control over the signals that are fed to the brain. Technology has brought many of these reality-hacking possibilities to life.

Will tampering with the boundaries between the land of the virtual and what we have come to know as reality lead to dark new outcomes (like losing touch with reality and meaningful human connections) or to beautiful, positive opportunities (expanding our imaginations into limitless possibilities)? As with every advanced innovation, the answer here is *both*. So although our steps with such a psychologically powerful tool need to be cautious, the wonders of this technology come in the form of harnessing our human ability to dream. Allowing us to bring fantasy to life. Providing us a means to journey into the imaginings of others.

We are creating new worlds.

Beautiful and functional spaces in the land of the digital.

Where you can go anywhere.

Where you can experience anything.

Parallel dimensions.

Where the real and the virtual become intertwined.

17

CREATING THE PORTAL

MAKING THE UNREAL REAL

To get to the virtual world, to view it, to interact with it – we always require a *portal* to this parallel dimension. That portal can, for Level 1 sensory stimulation (see Chapter 13), use display technology – screens, holographic glass, projections. These come in the form of eyewear like glasses or head-mounted displays (HMDs – devices worn on the forehead that look like big glasses, boxes harnessed to the head, or helmets) in more advanced forms of MR and VR. Smartphone screens have been used in simpler VR and AR applications. A next-level design we'll see soon will involve contact lenses we can discreetly use to view virtual elements within our environment.

Portals have evolved over the years to connect us with these realms of the imagination. Attempts to make us feel like we've been transported to another world date far into the past. The word 'panorama' was coined all the way back in 1792 by Irish painter Robert Barker, describing 360 degree mural paintings that would fill the viewer's vision and make them feel like they were in urban locations, cities and later nature, famous military battles and historical events. Stereoscopes turning 2D pictures into 3D images appeared as far back as 1838, when Charles Wheatstone adapted the binocular vision of humans to provide 3D perception. He showcased his stereoscope using two different drawings with a slightly shifted perspective (the first daguerreotype photograph, making photography a reproducible process, was still a year away). In the 1930s, a science fiction story by Stanley G. Weinbaum called 'Pygmalion's Spectacles' described a pair of goggles that gave the wearer an experience of a fictional world through holographics, touch, smell and taste. The mid-1950s saw filmmaker Morton Heilig create an arcade-style machine called the Sensorama that nearly engulfs the user for a multi-sensory experience. He further finessed this idea into a smaller design, inventing the first HMD, called the Telesphere Mask, patented in 1960. This didn't have interactivity or motion tracking to allow the person to look around, but the following year the Headsight was developed by two engineers, Comeau and Bryan, featuring a video screen for each eye and magnetic motion tracking. This was a very early precursor to the devices worn for VR today.

A range of other devices and designs followed over the next few decades. Along this line of thinking, 1982 saw the release of the film *Tron*, a visually stunning film for its time, in which a computer programmer ends up trapped inside a digital software world, trying to survive and escape this neon, foreign dimension where programs appear as people. A boom of interest and investment in VR came about in the late 1980s. Jaron Lanier, founder of the Visual Programming Lab, coined the term 'virtual reality' in 1987, and we started seeing large arcade devices made for public access in the early 1990s. The sci-fi thriller *Lawnmower Man*, released in 1992 and in part based on Jaron Lanier, introduced VR, albeit a pretty freaky story of it, to a general audience. Personal gaming brands soon released consumer VR headsets, including Sega and Nintendo, but these completely flopped. The headsets were cumbersome and the graphics were clunky.

I remember Dad letting me try a few early VR systems – headsets and 3D glasses. One of these systems, CrystalEyes by the company Stereographics, allowed you to view the computer screen in 3D. This company was founded by Lenny Lipton, who at the age of nineteen had written a poem that was adapted into the lyrics for 'Puff the Magic Dragon'. I loved all these pieces of technology despite many giving me headaches due to the jittery refresh rates. The time just wasn't right. The technology died out while the world focused on the rise of the internet and the dotcom boom. Imaginations had been sparked nonetheless. The reality–virtuality continuum concept was first introduced by engineering professor Paul Milgram in 1994. And then *The Matrix* by the Wachowskis hit cinemas in 1999. What a visionary film it was too, bringing together ideas of AI rising up, battles between humans and machines and the creation of a virtual world to imprison our minds – the Matrix. It took us into these brain-twisting ideas that we could potentially be living in a simulated reality and not even know it.

I definitely felt VR would one day return and was looking forward to it for a range of reasons. In senior high school (2001) I became instantly hooked when playing the game *Halo: Combat Evolved* on Microsoft's Xbox console. Set in the twenty-sixth century, *Halo* is a first-person shooter game where your character is a cybernetically enhanced super-soldier called Master Chief. You are assisted by an AI called Cortana, and the relationship between you both is a crucial part of the *Halo* game story, as effectively she becomes your guide and assistant throughout the journey. Cortana would go on to become the inspiration for Microsoft's intelligent personal assistant of the same name, first demonstrated in San Francisco in 2013.

While exploring and uncovering the secrets of the ring-shaped artificial world, you battle waves of various species of aliens. You also do a lot of running, ducking, dodging and jumping while wielding a wide range of weapons . . . well, that is, your character Master Chief does. In stark contrast, you as the player only really give your fingers a solid workout on the controller. This is what made me pretty obsessed with the ideas of when VR would one day make a return and what might be possible in the next wave. I loved daydreaming of when the time would come that we could go into a game, run around and be completely immersed, mentally and physically. In first year university, I started looking into using my engineering studies to potentially down the track create devices like an 'omnidirectional treadmill', a term that only became known to me years later. The idea was a treadmill that allows you to run while immersed in a game, not just in one direction like a conventional treadmill – but in any direction! I thought I'd get to really experience all that exercise my Master

Chief character was getting. He would run up and down hills, so I also wanted this treadmill to change in conjunction with the terrain. This would be a huge engineering challenge incorporating some robotic principles. I started planning for the prototype to form my undergraduate thesis. But then the diving accident happened and my trajectory changed for the rest of my degree.

At the time I started viewing other games and movies through a new lens. One of my favourite book sets is *The Lord of the Rings* by J.R.R. Tolkien and I loved the movies too. After watching the Battle of Helm's Deep in the second film, *The Two Towers*, I wanted to be immersed in the battle. The film captures the feeling of the soldiers waiting to defend the fortress as Saruman's army of Uruk-hai (big beasty orc creatures) closes in. I dreamt of fighting the enemy as Aragorn, feeling the weight of the sword in my hands and being immersed in the action and atmosphere.

I wondered if this would one day be possible.
To find yourself in the shoes of fictional characters.
To see the world through their eyes.
To be them.

Some years later, when I was working on my mind-controlled smart wheelchair, TIM, I was going through many frustrations with the two camera systems (stereoscopic and spherical vision) attached to the wheelchair. Whenever I connected them both to the tablet PC at the same time, it would crash and give me the Blue Screen of Death (when a Windows system crashes, the system error message appears on a full-screen blue background). This was so frustrating and required persistence and problem-solving. One night before heading to bed after a long day of failing to solve this problem, I was hit with a vision. It was a ball of cameras facing outwards, which would solve all my problems if it existed, and actually make designing the vision for the wheelchair much easier. It would be able to see a full 360 degrees in 3D vision. As I drew up my plans and researched the types of cameras I would build into this *camera ball*, a new revelation hit me: if these types of cameras were possible, then entire environments like rooms in houses would soon be able to be filmed, creating 3D representations for VR.

I instantly put together a pitch in 2008 and, not really thinking like an entrepreneur back then, tried to give the idea away to those who had time and funding and business acumen. I told them that this camera ball with stereo overlap in each direction could be created to produce 360 degree 3D vision. We could film many places around the world for the day VR would return. When asked when this would occur, my consistent answer was, 'I don't know. Sometime soon.'

As it turned out, the technology required to bring back VR came packaged in our smartphones. A huge set of innovations brought a range of capabilities to these devices in our pockets, particularly a high-resolution small screen, an incredibly powerful CPU (with millions of times more computing power than NASA's Apollo 11 guidance computer – truly amazing for its time – used to put man on the moon in 1969) and an inertial measurement unit that tracks the angle and movement of the phone. With these features combined, you take a phone, stick it in a headset with a few lenses so your eyes can focus on the screen, and what you see is a virtual world, with the displays shifted between your individual eyes for 3D perception. You start facing one direction and as you turn your head, the device recognises its own movements and pans the images being displayed. This means you can look in any direction and it will show you what you'd be looking at if you were there. It becomes a neurological trick, providing an instant feeling of immersion. But this is the simplest form. Devices and features get much better from here.

VR resurfaced in 2012, heralding a new wave of extended realities. And this time around, it was here to stay. It began when a new HMD called the Rift was launched on Kickstarter (a crowdfunding website where potential buyers can back products and help get them to market). It was produced by the company Oculus, led by young entrepreneur Palmer Luckey. The campaign raised nearly US$2.5 million. Mark Zuckerberg had Facebook acquire Oculus for around US$2 billion in 2014 as he could see the potential in it all.

Of course, it didn't stop there. Further devices continued to be released, and these portals even led to a greater emergence of devices for AR. Even though the concept of creating holograms and interacting with them had been around for decades, the technology was finally catching up. Many researchers over the years attempted to develop various forms of hologram technology, originally inspired by the famous scene in the 1977 film *Star Wars* (later subtitled *A New Hope*) where Princess Leia appears as a hologram, projected from R2-D2, with a message pleading for help from Obi-Wan Kenobi. But holograms were too expensive and difficult to produce effectively, so the simple idea eventually emerged that if we can *perceive* holograms in our environment, then that's achieving the vision. This is a basic idea of AR; we just need a technological portal to visualise these digital holographic objects in the real world.

The first time AR truly spread around the globe was through the smartphone game *Pokémon GO*. A virtual game land was interlaid with the real world, turning real-world locations into places *for catching Pokémon, for obtaining game resources and even as battlegrounds*. The AR portal was your smartphone. The app would open the camera so that when you viewed the real environment through the screen you could see the virtual world and *characters would appear* in your vicinity. It changed human behaviour across the world. My brother Alex drove home from an evening shift at the hospital one night to find that the roads to his house were blocked by people – at 1.30 am! – because the park across the road from his house featured *three PokéStops* within very close proximity. It was like a 24/7 rave, with people constantly there *picking up resources quickly* without having to travel. The thing about a lot of real people playing the game in the one location is it *brings many more virtual Pokémon to the area*, which in turn brings more people who just *gotta catch 'em all*!

With the release of various HMDs for interactive AR, and simpler glasses and even contact lenses on the way (let alone sensory-hacking levels beyond Level 1), the technology will increasingly infiltrate many areas of our lives. Cameras and other device sensors allow the real world to be mapped, intertwining the two realms and allowing them to effect change in each another. The ability to *summon interactive holograms* into the world around you is a powerful thing, and when the time comes (and the technology is simple and accessible enough), many will choose to use these devices instead of smartphones. They'll be increasingly utilised in areas such as education, to allow teachers to bring up holographic educational material the class can see and interact with; hospitals and surgery, where surgeons will have access to charts, AI assistance and remote colleagues around them; collaborative projects and telepresence; entertainment and games; and general computing and productivity. It will also open up a range of virtual controls, connecting individuals to the smart devices around them in their homes and world, even assisting people with disability to independently access computing and their connected environment. With many opportunities yet to be seen, devices that act as portals between real and virtual worlds are continuing to emerge . . . and are changing our reality forever.

Portals have been built,
to dimensions we have created,
to access lands of the virtual,
to experience the depths of the imagination.
So how deep does this rabbit hole really go?

18

VIRTUALLY FACING
YOUR FEARS

VR IN PHOBIA THERAPY

I managed to get my hands on the first Oculus Rift Developer Kit (DK1) some months after its public release in 2013. I tested it out and was instantly convinced that the new future of VR we'd been waiting for had finally arrived.

I tried the basic experience that came packaged with the DK1 software downloads, called *Tuscany*. You put on the headset and *find yourself outside a small Tuscan house on a tiny island high up off the water. The display and fairly low resolution make it look like you're looking through a screen door. The real selling point is the fact you can look in any direction, instantly making it quite immersive. You can hear the sound of birds and the waves down on the rocks. You can use a keyboard to move yourself around in this little world,* into the house and to the edges of the yard overlooking the water.

It felt like this device was unleashing my imagination and stretching my mind's limits. I was downloading a range of experiences and at the same time learning to create. I introduced myself to the effects of VR horror with a mate, Nathan, sitting beside me on my couch with the DK1 running off a laptop playing the VR *Horror Tribute* experience.

I found myself in the centre of a wide couch in a lounge room, and looking around, there was a TV in front of me playing the news, a hallway to the left of me, a low bookshelf to the right, and a lot of space behind me between the back of the couch and the wall. Nathan sat and talked to me but couldn't see what was going on in the headset. While I explained what I could see, *the virtual TV in front of me turned off. I looked around. It stayed off for a while, then suddenly turned back on, displaying static, before showing a familiar scene from the horror film* The Ring, *an empty woodland in flickering greyscale, with a water well in the middle. The dead girl from the film climbed out of the well and began to walk towards the screen. My heart started racing. She almost got to the screen and then the TV turned off again. I sat there staring intently at the TV. It didn't turn back on.* 'What's happening?' asked Nathan following my extended silence.

'The TV's gone black. Nothing's happening. I'm sure it's going to turn back on and she'll come out through the screen . . .' I waited a little longer. 'But nothing's happening.'

Then a chilling shockwave spread throughout my body as the moment of realisation struck that there was a fundamental difference between horror in VR and horror in any other digital medium we're accustomed to. When watching a film on a screen or in a cinema you always know you can turn away. *In VR, a fundamental difference is if you turn around, those nightmarish entities could now actually be right behind you.*

'Oh shhhiiii . . .'

'What is it?'

'I think she's standing behind me . . .' And before I let the fear factor really sink in, I quickly turned around . . .

Nothing.

I looked around the room. Still nothing. 'I think it's frozen.' *I looked around a few more times* and just before taking the headset off *I looked down at the other side of the couch. There, in the previously empty space next to me now sat the creepy horror film doll Chucky, watching the TV that had started playing static again.*

'It's that Chucky doll . . .' and before I could finish that sentence, he suddenly turned and looked straight up at me! I let out a loud yell and that was as far as I was going in this experience. The aim was to freak me out and it worked.

Moving on to experiences that can demonstrate other psychological effects of VR, like our perception of time, I downloaded an application called *Sightline: The Chair*, which quickly became a favourite for me. In this experience you sit on a chair, and *as you look around things change in the directions you're not looking. You find yourself being teleported from one place to another as the world changes around you, from fields of grass to city streets to small rooms to floating through space to sitting precariously high up on construction girders.* This is just the glimpse I've been looking for into what VR can do. *Time strangely seems difficult to keep track of, since the mind isn't used to suddenly teleporting between so many places.* I emerge feeling like this experience was about fifteen minutes long, when it has actually only been five minutes.

I try *The Chair* on my maternal grandmother, Shirley. Grandma is a beautiful, very miniature kind of human, reaching a whole 1.5 metres off the ground – I'm pretty certain my height had absolutely nothing to do with her. My mum often reminds her of the saying 'Good things come in small packages', to which she responds, 'Well, I must be pretty good then.' She's the most amazing grandmother I could have asked for and has been a constant guide in my life since the start, always supportive and loving.

After her experience with *The Chair*, Grandma also felt like her mind had just gone on a long journey in a matter of minutes. She really enjoyed it and asks me if VR could help her get over her fear of spiders. Ignoring the previous jump scares from sudden shocks in VR, I said slowly, while trying to think it through, 'I don't think so . . . because you'd know it's not real.'

Now, just for context, my grandmother's arachnophobia is greater than that of most people. Let me illustrate with a story I honestly can't believe happened to her, of all people. One day Mum had called her to say she was coming over to visit soon. During that time, Grandma pulled down the window roller blinds and a huntsman spider fell off one of them, landing directly on her face! About an hour later, my mum arrived and found Grandma standing there, frozen, still looking up. She had become completely petrified when the spider had landed on her face. By this time it had already crawled down her body and was long gone, while poor Grandma was still an upward-looking miniature human statue.

Despite my initial scepticism about the usefulness of VR for phobias, my own less intense arachnophobia would soon be put to the challenge. After a few months I try out an experience where *I watch a virtual spider crawling up my arm. It gets as far as my shoulder before I leap up,* throw the headset off my face and frantically shake my hands through my hair. My mind has made it real for me, and my reaction feels real too.

So, what does this mean for us? Well, if our minds allow a similar response in a VR experience to if it was happening in real life, then the level of immersion is clearly a very convincing thing. I started to think that this might just be a tool for facing one's fears and gradually getting over them through VR exposure therapy. The idea is simple: if you're exposed to something you fear over and over, your mind stops giving the same response and eventually becomes desensitised.

During the filming of my ABC Catalyst documentary *Meet the Avatars*, I try out exposure therapy using 360 degree filmed VR with therapist Corrie Ackland of the Sydney Phobia Clinic, who brings a headset and a St Andrew's Cross spider called Webster in a cubic soft-mesh cage. She shows me the cube and when I lean down and see Webster inside, I immediately have a strange kind of dry retch reflex. Corrie talks me through the idea of phobias and our responses, and works it through with me about what I feel and why I might be having this reaction.

The idea is that trying out exposure to Webster in 360 degree filmed video in the VR headset before the real thing will allow me to face my fear for an extended period, until it no longer gives me the same near-heaving level of response. 'Exposure therapy is basically addressing the avoidance of a strong maintainer of a lot of these fears,' Corrie says. 'Particularly spiders, relatively easy to avoid, run away from, get other people to deal with, that sort of thing.' She continues calmly, 'So shall we get some practice with his movement, maybe in virtual reality?'

I'm someone who pushes my own limits quite often, so I'm surprised to find this as confronting and fear-inducing as I do. My phobia was initially triggered by being bitten by a spider when I was young. A phobia I was under the impression was now largely gone. I can tell Corrie is wanting me to put this spider on my hand. No part of me is wanting to let her. As she hands me the headset, I can already feel my heart rate going up.

I slip the headset over my face. *There's a basic room with filmed arms in front of me that aren't mine.* This brings up an immediate urge to move along with *what the hands are doing.* It's an odd sensation when *your virtual arms and hands are doing something* you're not. *They unzip the cube cover, fold over the flap to open up the top side, and the right hand reaches in. Next thing, Webster is walking along the hands as the right transfers him to the left and back again, by blocking the direction he's walking in.* Although we don't have the amount of time you'd normally put to this form of exposure therapy, I feel I'm already starting to get used to *Webster's movements* after five minutes in the headset. Corrie asks me what I'm worried about and I tell her I'm not entirely sure – if it's not dangerous it doesn't really make sense. I explain that not knowing how he moves or how quickly is a factor, and the unpredictable nature of Webster is getting to me, as I'd hate to end up with him running up my arm and into my shirt. All those fears are eased now by *seeing how Webster moves in the headset, keeping a fairly consistent pace,* as well as Corrie talking me through how he only ever goes forward in the direction he's facing.

She asks me if I'm ready to give him a go in real life. I feel if I'm ever going to be ready, now is the time, before my mind gets the better of me. She takes him out. I watch him run up her arm as I start building myself up to take over. 'Yeah, give me a second.' I try to convince myself I'm not fearing this.

Corrie says, 'There's only one way you're going to break that.'

Another brief hesitation. I lean back, take a deep breath, lean forward again and say, 'Come on, let's go,' as I put my hand out. Webster runs straight onto it. My heart races. I feel discomfort. A few moments pass and I think like I've been holding my breath. Then the fear slowly melts away. Webster is admittedly starting to seem bizarrely cute.

Most of the time, facing our fears can help us get over them, and exposure through technology such as VR can aid in the process, especially if facing our fears in reality might pose a legitimate danger. When the thing we fear is dangerous, these fears are completely rational, but cultivating a calmer, regulated adrenaline response could potentially be more useful when actually faced with such situations – as opposed to the very common scream-and-freak-out phobia responses.

To see how others deal with this concept in the documentary, I recruit a group of volunteers. We're going to attempt to tackle their fear of heights. In this experience, the user starts by *going up a building lift from street level. When the doors open they find themselves high up on the edge of a skyscraper, with a plank of wood protruding out from the edge. Although it looks clearly computer generated, the experience really sells the height factor to the player's brain – straight out are the tops of other nearby skyrises, light clouds floating, birds flying around lower than the plank, and in addition to the vision it comes complete with realistic sounds of traffic below when you look down, gusts of wind when you step out and a helicopter making the odd flyby. The objective is to walk the plank and not fall off, otherwise you'll plummet to the city below.* Our first few participants manage to face their fears and *walk the plank* without too strong a reaction. But one volunteer, Gerry, stands back and laughs.

He tells me, 'They look a bit silly, going "Oooh, it's so scary."' He adds some nice hand-waving gestures to accompany his sceptical comments.

He doesn't feel like this will work on him, so I say, 'You haven't tried any VR before, have you?'

'Nuh, but surely you know it's know it's not real,' he says.

'And you're afraid of heights?'

'I couldn't think of anything worse.'

I set Gerry up with a headset, headphones and heart-rate monitor. He stands there for a moment, completely calm, until *the lift doors open and Gerry looks down to the city below. He is clearly immediately gripped with fear* and we watch a real-time feed of his heart rate start climbing.

'Are you willing to walk out onto this plank?' I ask.

He tries to reason himself through it logically. 'I know I'm standing on the ground. I know where I am. I don't think I can do this.'

I offer a little encouragement. 'Yes, you can. Yes, you can. Take your time.'

He swears under his breath nervously, looking around, his breathing getting heavier. 'I'm seeing my worst nightmare.'

Gerry's heart rate spikes at 132 beats per minute, almost twice his resting heart rate, and he hasn't even moved. His breathing gets even faster. He starts shaking and sweating profusely.

'I know this isn't real,' he tells me, still trying to reason his way through this.

'That's good. Tell yourself it's not real.'

'It's not real. It's not real.' He stops and starts stepping backwards. 'Oh man, I can hear a helicopter,' he says as a virtual one flies past him. He swears a little more.

I go over and *guide him out along the plank*, then realise he's not reacting. 'Do you want to open your eyes?'

'No. No I really don't,' he shoots back as I start laughing and wait for him at the starting point.

He awkwardly turns himself around and leaps along the plank and back into the lift, back to his starting point, where I catch him so he doesn't fall over or run into our equipment. I'm incapable of containing my laughter at this point.

Taking the headset off Gerry, I tell him, 'You did so well,' and he thanks me for the experience.

I continue: 'Everything was telling your brain that that was real. I mean, we could see it in your heart rate, we could see it in your response and how much you're sweating. Your response is real.'

'I can see it looks like a cartoon. It looks like a game. I know I'm part of a game, but my brain . . . was so convinced,' he explains as he turns to me wide-eyed and completely converted.

The power of this tool has quickly become apparent. If it can elicit strong emotions like fear and provoke similar reactions to actually being in the situation, this could have some seriously negative effects – a potential for heart attacks, say, or some form of trauma. But on the flip side, there may be many positive effects too, that were previously unimagined and impossible to achieve.

A powerful tool to elicit emotions,
VR takes fear to a whole new level.
We haven't yet had the time to realise
the true effects of this.
Immersive horror experiences may traumatise,
or they may desensitise,
and there is still so much to learn about
this human emotion.
But one thing we know
is that VR allows us to face our demons.

19
BUILDING OUR
VIRTUAL WORLDS

LIFE-IMPROVING POSSIBILITIES
OF EXTENDED REALITY

The devices already available to the average consumer for extended-reality experiences mostly have good-quality visual stimulation, high-quality hearing, moderate-at-best touch, the very starting points of smell and almost non-existent taste stimulation. So as advanced as the technology may seem, we're still improving these Level 1 sensory-hacking technologies, and are still only in the very early days of its potential. But oh, potential there is.

Many areas of real estate have made big steps towards utilising the power of VR – to *take people through the layout of homes, or even in the design of new buildings, to see, walk through and alter before it's even constructed.* Training is changing in all kinds of regular work and in such fields as the military, veterinary surgery and the US space program, as a cost-effective approach for astronauts preparing for space missions. VR can provide realistic simulations of environments and take a soon-to-be astronaut through *accurate representations of entire stations like those planned for Mars, carrying out tasks and procedures, simulating emergencies and operating rovers.*

Some previously unimagined possibilities have surfaced in exciting areas with positive impact, such as rehabilitation. A range of research has yielded promising results in using VR to assist in rehabilitation for disabilities such as spinal cord injuries. One study from Brazil, resulting from international research collaborations, involved eight patients who were completely paralysed from spinal cord injury. None of them had improved with previous types of rehabilitation, including walking on a treadmill with the assistance of an exoskeleton. However, this study progressed further to fuse a range of approaches, combining immersive VR training with a lower-limb exoskeleton for walking on a treadmill. The big difference from previous approaches is that instead of the patient seeing that they were being moved by the equipment, inside the headset they saw scenes like *their legs independently walking on grass.* The power of this imagery can trick the mind into believing it has independent control over the body. This is where the magic happens. Following this program, all eight patients showed signs of improvement, including regaining sensations of touch and pain and even voluntary muscle function and movement.

What's more exciting about the prospect of such research is that similar examples are being trialled in Australia. I wanted to see this during *Meet the Avatars,* so I travelled to Newcastle to meet virtual therapist Rohan (Ro) O'Reilly, who has been testing these ideas out with his patients. The day I visit, Angus is trying out Ro's rehab techniques. Angus, a cyclist, had a horrible accident five years earlier when a car drove across his path, leaving him paraplegic from a complete C7 spinal cord injury. What do we know about complete severing of the spinal cord? No signals can travel down or back up across the site of injury in the spine, hence complete loss of movement and feeling. Well, that's what we think we know. This research is really starting to challenge that fact, so I'm keen to see what they've achieved.

Ro and his team set Angus up with a range of equipment. Angus sits in a chair while a functional electrical stimulation (FES) system (straps with electrodes that send low-energy electrical pulses to generate body movements) feeds electric signals to the muscles in Angus's legs to make them move. Ro's assistants help with this exercise to provide smooth movements of Angus's legs. But Angus doesn't see any of this happening. Instead, he is using a Vive VR device (by HTC), wearing a headset and holding a controller in each hand. He's experiencing *holding pickaxes, throwing them into the wall of ice as he climbs the side of Mount Everest with a group of climbers, independently moving his entire body.*

Angus has been undergoing this type of therapy for about six months, and it seems something truly incredible is happening. I believe it to be due to the power of optimism, combined with a mixture of the placebo effect – as the brain is being temporarily tricked into feeling like he is independently controlling his body – and neuroplasticity. What we see is Angus is starting to feel, and even move. He shows me that he can now, with great effort, move his foot, and he is constantly improving bit by bit. Then he tells me he is going to show me something that was thought to be impossible. He tells me he is going to stand.

In another, larger room, Ro loops a harness around Angus and connects it to a hoist hanging above. The hoist isn't there to lift him up, just to stop him from falling. With an assistant holding his feet down, Angus takes hold of Ro's arms as support. He has the FES system set up to help straighten his legs. Despite all this, actually getting upright is all down to Angus. He has a go, but doesn't get very far before he sits back down. Another try. Same result. It's actually making it more real for me that his first few attempts aren't working. He's starting to worry he is too fatigued to show me this feat he has been so excited to display. He takes a moment to relax and regain his strength. A big breath, and he gets ready to try again. Ro rolls up his sleeves instantly sparking Angus's humour: 'Show your guns!' He grabs hold and I can see the wave of determination move across his face, the sort I've seen before in Riley. A slow exhale to deepen his focus follows. He makes a move.

Slowly, he moves forward. Slowly, he starts rising. He passes the previous positions where his body failed. I can see the intense effort and energy being expended as he draws on every bit of his mind and body to push himself up towards a vertical position. He makes it! Standing upright, he expels a big breath, then looks Ro in the eyes. Ro says, 'Hey buddy,' as they both let out a small laugh of relief. I'm astounded at what I'm seeing. I feel like I'm witnessing a miracle.

Another therapeutic application of VR is in phantom limb pain, a perplexing phenomenon that occurs in most amputees. In no way a fun phenomenon, it's one that has baffled medical professionals since it was first discovered, with the term 'phantom limb' being coined by physician Silas Weir Mitchell in 1871. What happens is sometimes amputees will start to feel pain in the hand and fingers, or legs and toes – of the limb no longer present! Some describe sensations of itchiness, cramping and numbing. Until recently, *mirror-box therapy* was used as a method of treatment for this. Basically, a set of mirrors was set up so that the person experiencing the pain, say with an amputated right arm, would hold out their left arm, and from their perspective see that they had two arms – the right a mirrored reflection of the left. For many who tried this therapy, once they perceived both arms the pain intensity and duration would often reduce, as if the brain was saying, 'Oh, there you are, limb. I've missed you.'

This can be created even more effectively in VR and AR. With the use of motion sensors to track the movements of the limbs and mirror those to the other side, or biological sensors to decode the signals the brain is attempting to send to the phantom limb, the person perceives both limbs are present and that they have control over both. VR and AR hold much promise in being some of the most effective, low-cost methods of reducing this type of mysterious pain.

I've also loved seeing VR apps for needle injections. Getting a vaccine injection can be terrifying for many children (and adults too), so these apps are made to alter a child's sense of the experience, calm them and alleviate the pain. In one called VR Vaccine, a child has a VR headset on, and in the experience *they are a hero, about to save a kingdom from bad guys.* The timing of the nurse cleaning the skin with alcohol and administering the needle injection is matched step-by-step with the child's experience of *being given ice and fire superpowers to use on the virtual enemies*. Transforming their sensations from the procedure alleviates their experience of pain. Awesome!

Again I feel inspired by the creative uses of these advancing technological tools that are quickly making the impossible possible. A new world is upon us and the younger generations coming through are growing up in a very different one from those before them. They don't know a world without all these connected devices and smart technologies, which are still somewhat novel and sometimes confusing to the generations their senior.

I learnt about these *smart tech natives* first-hand in 2014 when I was a judge for a national competition for school students to put together their own technology applications, completely self-directed. Arriving at the venue, I see table displays set up all around. A great deal of time and effort has been put into each stall and I now want to make my way to see every one. I stroll over to the primary school section and instantly notice one table has an Oculus Rift DK2 VR set-up on display. This is the first time I've seen the students use VR at one of these events. There are two girls in Years 5 and 6, ten and eleven years old, and their story is just awesome. They both have parents who are doctors so they've learnt about things like how the heart works and heart disease. They tell me about how other kids in their class don't know about these, inspiring them to create an application to assist in their learning in a much more interesting way than the traditional approach.

'Kids in our class don't really like learning just by reading from a book,' the eleven-year-old explains.

'Yeah, we're a very multi-modal generation,' says the ten-year-old.

I just stare blankly at her for a bit at this point. 'No kidding,' I muster up as I'm working through her words, thinking about how I definitely did not have that type of vocabulary at their age.

She goes on, 'So we turned learning about the heart into a fun program in virtual reality.' They found a free tool online called Blender, which allowed them to create their own 3D digital models – a heart, valves, blood cells – and then bring them all together into another tool with free personal access, called Unity. This is one of the most popular game-development engines, with a huge range of uses, including development for VR programs. (Unity versus another amazing engine known as Unreal Engine has been a long-standing battle of the giants, comparable to Microsoft versus Apple.) The girls built their heart experience for the Oculus Rift DK2, packaging up the software and running it on the device. What happened next was brilliant.

I put the headset on and *found myself shrunk down, inside the heart, watching the blood cells fly past before flowing through the heart along with them. You never learn quite this well about how the heart works until you've taken a theme-park tour ride through it.*

I lift up the headset. 'Wow, it's like *Innerspace*! You've seen that movie, yeah?'

They look at me a bit blankly. No, of course they haven't. This project came right out of their own creativity. So well implemented and documented too; I find this astounding. And there are many other projects around the room that are just as amazing.

These experiences make me want to see more. A few months on I've been invited, by an old schoolmate, Alex, to Bondi Public School to speak and help with the judging on the finals of a Year 4 app competition. Yeah, these are all about nine- and ten-year-old kids, designing and building their first apps. I ask the teacher how she has made this happen and if she herself learnt to make apps before teaching her students.

'No, I've never made an app in my life,' she says. 'What I did was I asked the students to come up with an idea for an app. I told them if they like skateboards they can do it on skate-boards. I didn't mind what it was, just that it was personal to them and something they wanted to do.'

A simple yet effective approach to innovation. She told her kids that they would each have their own app, learn how to make it and collaborate and help each other out, because the project wouldn't end until everyone had their own app. Then at the end of it all, she made the most empowering suggestion she could possibly have made: 'You can teach me how you did it.'

They have built a range of apps for various smartphone and tablet systems, ranging from apps to aid in learning scientific concepts to helping track the neighbourhood movement of their pets. Most seemed to address social and environmental concepts and I love that so much. How collaborative and supportive they have all been is an approach that should be implemented often. It's an ideal mix of competition *and* collaboration.

And now, this girl presenting is really someone special. Full of confidence, she hops up to use PowerPoint in a cleverly constructed presentation. She has the makings of a true entre-preneur and immediately gives the impression that she has the inherent ability to sell anything! Nevertheless, she's clearly about to present something amazing.

'When we started this project I read a newspaper article that had on the front page the number of Australians who die from getting caught in rips at the beach each year,' she says. 'I thought, there should be an app for that! So I decided to make an augmented reality app for it.'

'Wow, all right,' I say as I think it through. Couldn't have predicted this is what I'd be seeing here. Your average person at this point in time hasn't even heard the term 'augmented reality'. Her app has a well-planned process: when you arrive at the beach, you start her app, it accesses the camera, you hold it up and, looking at the phone screen, scan the beach and water as if taking a video.

She goes on, 'I haven't quite got the image processing working yet to identify where the rips are—'

I respond jokingly, 'That's okay, I'll let that one pass,' thinking *I can't believe she even knows the terminology she's using here.*

She continues, 'But I can simulate it.'

And I say, 'Of course you can,' my inner self laughing and by this stage in complete awe.

'So when you've scanned the water, you aim the phone back at the water and look at it through the screen. Words appear above the water to say where it's safe to swim and where the rips are. But I did this much faster than I expected and I realised you'll only understand it if you can read English.'

This is where the extent of her thoughtfulness shines. She is still displaying outside-the-box thinking and realising her app can be scalable and more inclusive. She looked into universal design principles to find symbolism acknowledged across the globe. She has landed on implementing an app design that shows smiley faces where you can swim and big red X's where you can't. If you can't figure that one out, you're probably gonna drown!

She then brings up a slide displaying a smartphone image of her app's home screen. 'She grabs the mouse connected to the computer running her presentation. 'And this is how it works,' she says as she proceeds to click through her app menu and features.

I briefly interject: 'Whoa, hold up! Did you embed a smart-phone app emulator in your PowerPoint presentation?'

She looks back at me and, with oh so much sass, says, 'Yeah,' before turning back to the screen and continuing.

'I didn't know you could do that!' I say, as it sinks in that this technologist just got schooled by a nine-year-old. As if that isn't enough, she has already engaged her stakeholders, heading out to the beach and interviewing lifesavers and beachgoers, asking them if they would use the app and if they would like to see any other features in it. She finishes her presentation with interview videos and her resulting future implementations. Now also bear in mind, this was towards the end of 2014, a time when few adults had even started wrapping their heads around the coming age of extended-reality technology. This was just one example of a young mind's adaptability and true creativity.

Since this time, I've seen the space of extended-reality technologies steadily take off around the world. Adoption and general use comes and goes in waves, as do most types of new flourishing technology. I love seeing the evolution continue, but I still think about this time in 2014 when these youngsters, and others like them, were able to harness the technology early to implement and achieve some pretty audacious ideas. Ideas most adults would think too difficult to even start on. What I take from this experience is that we all remember what it was like to be young with a big imagination. Creativity is seemingly inherent in us, but many of our previous generations had a lot of that knocked out of them by the education systems of old. Even today, we're still seeing archaic schooling systems finally starting to adapt to the times – a long, ongoing, necessary process.

Maybe we can invite more of our own childhood imagination, full of limitless possibilities, back into our lives. That's why these kinds of initiatives are so inspiring. Guiding the emerging generations to hold on to their creativity in the first place. Fostering our young minds of today. Empowering our change-makers of tomorrow. Extended-reality technology to create our alternate realities will become the playground of the mind, and we'll see it increasingly assisting us in life and infiltrating our reality. So let's make it positive. Let's make it *human*.

A powerful tool for fabricating environments, eliciting emotion and creating experience, extended-reality technology can overrun our perception of both time and space.

20

SO PACK YOUR VIRTUAL BAGS FOR THE MATRIX?

CHOOSING THE VIRTUAL WORLD OVER REAL LIFE

Let's entertain this idea for a moment. You're offered a life of fantasy adventure. A place where you can do whatever you want to do. Be whoever you want to be. Where your mind can roam free. This was a vision portrayed by Steven Spielberg in the 2018 film *Ready Player One* (based on Ernest Cline's bestselling 2011 novel).

The movie's hero, Wade Watts, is a young guy whose life in Ohio 2045 is pretty bleak. The world seems like a giant junkyard on the brink of collapse. His virtual life is anything but. He puts on his headset and walks around on an omni-directional treadmill (yeah they really are awesome), but his *virtual avatar is an anime-style dude who drives around in a DeLorean time machine from* Back to the Future II *(with hover conversion and the addition of a* Knight Rider *KITT scanner on the front grill). Wade describes this virtual world, the OASIS, as 'a whole virtual universe. People come to the OASIS for all the things they can do. But they stay because of all the things they can be.'*

It's not as far-fetched as it sounds. Virtual worlds are getting more complex and massive. There are already some out there where people choose to spend much of their time, like *Minecraft, Second Life* and massively multiplayer online games and role-playing games, the best known being *World of Warcraft*. As the tech improves and the uptake grows, more dimensions indicative of the OASIS will appear – virtual worlds where large numbers of players, reaching into the millions, converge to *socialise, go on quests together and build a virtual life*. Some worlds have virtual economies, where you can buy virtual currency with real currency, and vice versa. So it is actually possible to *work within these places too. Communities, culture and social rules all play into many of these worlds.*

As the amount of content and users grow, there'll be many more things we can do in these alternate realities, and people will choose to spend more time *in them, even creating and maintaining virtual worlds from within.* Of course, there are many debates over the effects of long-term video game overuse. But what about from the less gaming, more virtual work and virtual life perspective? Could you see yourself *working in a space of unlimited size and possibility, unconstrained by the physics of our real world?* You could *spend your days under the ocean, in a high-tech lab or floating through space if that's what you desired. It could be a space where technology, education, tools, materials and textures are available on command.* As the balance shifts between the benefits of being in the real world versus the virtual world, so too will our time spent in each. As for the uptake of any other technology we can use daily, we each need to figure out our individual balance. We must continually learn what works for us, what's good for our own minds and what sort of mindfulness practices we bring into our lives to balance everything else out – such as diet, exercise and relaxation. In saying that, VR has the potential to assist in some ways here too. Not as a replacement, rather as another pathway we can take.

I felt the calming capabilities of VR in 2016. *Becoming Super-human* is premiering in a few nights' time on the ABC, which is also the night before TEDxSydney, and I'm presenting on stage first thing in the morning. My thoughts are spinning around as to whether we did Riley and the whole adventure justice in the post-production of the documentary. I'm not sure what viewers will think. We're also cutting it really fine with the technology for the TED event and are in danger of not getting the tech working in time. I'm usually fairly level-headed, but right now I'm tired and sleep-deprived. I can feel my worries weighing in heavily on my mental state so I drive over to my parents' place in an attempt to calm myself.

Mum and Dad are relaxing on the couch watching a movie. After chatting for a short while I let them get back to watching as I sit by myself in the living room, not sure what to do. This feeling is so heavy I'm finding it hard to breathe. I just want my mind to escape and roam free. To just chill.

I grab my laptop and my VR system out of the car. After

setting up and clearing some floor space in the living room, I launch Tilt Brush, an app where *you can paint in 3D* by moving around and using your controllers – which you *see as paintbrushes and holographic paint palettes. You have a large range of art tools – from standard paint to glowing trails, snow, fire, smoke and almost anything else you could think of.* I plonk myself down on the carpet and *select a night-time desert environment with mountains in the far distance surrounding flat land and a bright multi-coloured galaxy across the sky.*

I grab the fire tool and draw a fireball in reds, oranges and yellows, then surround it with earthy colours and textures and grass to form a mini glowing volcano. I create sparkling embers to flow up and out from the fire, glittering all around me, and throw in a few puffs of smoke from the flames too. Then I just lie down and relax, with the embers sparkling around me as I look up to the stars. I feel like I'm away from everything. Nothing but the mountains, this dark desert and starry night sky, my mini fire mountain and me. My feeling of anxiety soon seeps away. I feel calm. I feel ready for the next few days.

Since figuring out that this was a cool space for the mind, I've used VR for mindfulness a number of times. The 3D-painting part itself, not just the outcome, is also therapeutic. Then there are battle games, role-playing games, sensory experiences, stories, shooters and something I find particularly interesting – archery. Mostly because I can be standing there holding two controllers of the same size and weight – but *when I look down and see a bow in one hand and an arrow in the other. Strangely, my mind is telling me the bow is heavier, as it would be. When I put the arrow into the bow and draw back,* the controller

I see as the arrow vibrates lightly while *I'm drawing back. I can even look at it, think it through. My mind is telling me that I'm feeling tension and that this arrow is definitely connected into the bow* when the two controllers I'm holding are not touching at all. It's amazing how much your brain can make the virtual experience real.

VR has also been labelled an *'empathy* machine'. It's definitely a tool that can elicit empathy to a greater degree than most other forms of media, mostly due to the immersion and the feeling of *seeing the world through the eyes of another being.* This view can be computer-created or filmed in 360 degrees as if directly through that other set of eyes or from a fly-on-the-wall perspective. It has also been achieved through real-time mirroring of movements by a pair of people wearing headsets with their *vision switched, so they can in effect, be in the body of each other.* These types of immersion can provide an insight into the life of another, human or otherwise – provoking subjectivity, sociality, sympathy, perspective and even levels of empathy. Technology is not at all necessary to stimulate empathy, but done well, it can provide a powerful and scalable means of doing so. Whatever the approach, the world would be a much better place if we could understand each other more. It all comes down to human connection.

As I'm writing this I'm working with my Psykinetic team on creating our own form of AR Wizard's Chess like in *Harry Potter*, with inspirations from *Star Wars* and a computer game from when I was young called *Battle Chess.* Only here you have the options of hand gesture control, voice commands or eye control to help make it more inclusive, allowing a means of playing for pretty much any level of physical ability. You can play wearing the headset and *make the board and characters as large as you like. When you make the moves they come to life, fighting each other when they take an opponent's piece.* This is the power of extended-reality technology for bringing our wild ideas to life.

The possibilities really will continue to explode – in education, data visualisation and analysis, medicine, team work, psychology, telepresence, rapid prototyping, designing to explore difficult-to-reach places or building the impossible, simulating new environments and cities, creating multiple parallel dimensions interlaid with the real world, shopping, work, social activities and storing memories of experiences and loved ones. Even further into the future, perhaps we'll create new worlds, new lives and *porting of the consciousness.* Once it has been discovered what the consciousness is and where it is held, humans will find ways to move it, upload and download it and port it. We will one day be able to move beyond our Level 1 sensory stimulation (see Chapter 13) with extended reality, to perhaps as far as Level 3 sensory injection and then, imaginatively, Level 4 posthuman sensing, both of which will allow us the option to port our consciousness into avatars and experience pure virtual worlds.

The extreme, intense and controversially sci-fi ideas that arise here are those of technological immortality. What if we had the option to leave the confines of our biology and live on through a virtual world of endless possibility? Live without the need to consume a great deal of real-world resources and instead become one with our own technological creations?

In *The Matrix*, the character *Cypher strikes a deal with the agents of the Matrix in the hope of being reinserted into it, back into the virtual world in which he once lived. He says 'Ignorance is bliss' and wants to remember nothing of his time in the real world.*

Perhaps we won't want to live a virtual life in ignorance. What if we *choose* to retain full memory of our real lives and *be inserted into a virtual life, free from the struggles of the real world? How different would that be? Experience, connection, adventure – would we still find purpose in it? Societal constructs would be unrecognisably different,* but it's interesting to entertain the ideas. Even mixing these virtual lives with connection to the real world, using sensors and real-world devices, could open up new opportunities. With such possibilities, and through technology, we could travel out of our own solar system, out of our own galaxy, in ways our biology never would have been able to withstand. We could send out robotic space probes and beam forth our virtual consciousness, just to explore and experience the stars.

Now I know this sounds like the freaky fields of science fiction, and maybe it is. But I like to imagine rather than just let the initial fear response kick in. I prefer instead to take my mind towards the seemingly impossible limits, raise conversations and work through both the positives and the negatives. The fact is, anything is possible, so where do we want to steer these technologies? Depending on your own beliefs – spiritual, religious or otherwise – if uploading your consciousness and transcending your biological life into a virtual one were the only option, would you choose it over death? I think

I would in this situation. The hypotheticals are worth thinking about, so that when the seemingly impossible advancements arrive we're not caught off guard as a species. Having already pondered the possibilities, we'll know where we want to take them.

From the depths of the human mind
and the power of human endeavour,
we dare to dream,
and we find courage to act.
Every human-made feat
started as an unimagined impossibility,
sparked into an imagined impossibility,
transformed into an imagined possibility,
and eventually became reality.
So let's continue daring to dream.

PART V

DREAM OF HUMANS IN THE CLOUD

Just when you thought your mind
was full of enough wild stuff

21

THE DREAM AND THE GREAT CONVERGENCE

CAN WE PRESERVE THE MEMORY OF LOVED ONES?

A massively profound and impactful idea has just hit me like a freight train. It's early 2014. I'm visiting my parents, sitting with Mum watching a movie. But my mind is elsewhere. I've been thinking about the potential of these VR devices I've been playing with and of the emerging cloud services, which can provide data storage and supercomputing power when needed, along with other IT resources and applications such as advancing AI software.

My eyes start welling up. Not because of how mind-opening VR is, how game-changing cloud technology is or how powerful AI is. I'm feeling like this because I've just joined the dots and realised the weird thing I'm imagining is actually possible given today's technology. Many of the most ambitious ideas can be made reality through the *convergence* of a number of our advancing fields, bringing them together like super-solution building blocks. I get Mum's attention, mute the TV and try to explain what's whirling around in my head, through tears and goosebumps.

'I'm just thinking about how wonderful you are. How much, well, one day, when I have kids, I want them to know you how you are now. I want their kids and their grandkids and future generations in our family to know you. To know Dad. I'm just . . . thinking . . . there may be a way that these new things, new tech I've been looking into, could help make this happen. I mean . . . I think it might be possible to create something. It seems a bit ridiculous, but I know I could create a virtual copy of you. As you are . . . and I would want that for future gener-ations. I . . . uh . . . I'm just not sure how I feel about it. I just want to talk it through.'

In her warm, calming voice, Mum says, 'It's okay. What do you mean a copy?'

'I mean . . . Well, let's think about it as if the technology existed when Grandpa was alive. You talk about him a lot. It's always nice to hear – he seemed like such a great guy. You said I remind you of him?' (My maternal grandfather Richard passed away just before I turned two. In our family albums there are photos of him holding me as a baby and putting his big hat on me, completely engulfing my head. Unfortunately, he passed away before the triplets were born.)

'You do,' Mum says, smiling. 'I see my father in the way you walk. You have his smile and mannerisms . . . even . . . even your sense of humour.' She starts laughing. 'You stir the possum like he did. The way you sometimes irritate the triplets. So many things you do I just go, yep, that's Dad.'

That makes me laugh too. It's nice to know that I've taken on part of Grandpa's personality. I pause again and find myself staring down as I gather these thoughts. 'So. If this had existed then and I'd been able to create a virtual copy of Grandpa, I would want it. I'd love to be able to go into the virtual world and talk with him. See his face and his gestures and mannerisms, hear his voice and just . . . be there with him. It would be like a modern version of visiting someone's grave and sharing stories with them. How would you feel if that were possible?'

Mum starts tearing up now. Responding is clearly very difficult for her. 'I'm not sure how I feel about it. I think I would definitely want that for you and future generations in the family . . . to know Dad. But I wouldn't want it for me. Because . . . I'd be reliving the loss over and over.'

For a moment I feel horrible for even coming up with this idea. It breaks my heart to see my mother upset. Then I remember something I've come to know all too well. That if we let fear get in the way, be it fear of failure, fear of how people may respond, or fear of not living up to the expectations we set ourselves, then we'll never explore or discover anything. Instead, I believe we can take on board the thoughts of the people around us and incorporate their responses in planning and creation. Knowing that this particular design could provoke an adverse reaction means that I should approach the idea with caution. While the application of technology to create a digital avatar of my grandfather could have a negative – even traumatic – effect on Mum, it could also have a positive – even beautiful – effect on me, my siblings and future generations in our family. This is why we must approach all bold steps with respect and conscious design.

Conscious design. Yes, we must remain cognisant of the impact of what we create. But is the idea to preserve the memory of people alive today in this way even possible? I have no doubt it is. Cameras are advancing. So too are the algorithms that can do tricky things with camera footage. AI will come in useful for *natural language processing*, which can segment and figure out the words a subject is asking the digital avatar, while chatbot-style AI can recognise the intention behind the question and match it up with an answer from a database. Any form of extended-reality device could be utilised, either to bring the interactive avatar into the real world or to go into the virtual world and *meet the avatar there*. Finally, cloud connection, storage and process-ing will ensure the avatar is never lost and is updated with the latest advancements as they come, such as improvements in realism and interactivity. Yes, it is possible. We have the technol-ogy to create a virtual avatar of your living hero, or of someone who inspires you, or of your family members . . . or of you. How would this work? Does the avatar need to be directly controlled or can the conversations happen when the human counterpart is no longer around? Or can the avatar even keep learning for itself and go beyond what was originally captured?

The dream would be to find if there is beauty in capturing an individual and allowing their friends and family to interact with them, even beyond death. Working through the various possibil-ities has led me to develop this simple framework for classifying our avatars, real or virtual, as they will become increasingly commonplace as we move into the future.

A HUMAN'S GUIDE TO FOUR KEY AVATAR CLASSES (AC-4)

Class I: Transient avatars

Those you can control (mostly in real-time), often temporarily, while playing video games, through robotic telepresence or through virtual experiences

Class II: Virtual twinning avatars

A data representation of you, for purposes such as medical AI that can analyse, predict and improve your real health

Class III: Spatiotemporal avatars

A representation of you caught in space and time, for interactive, historical and memory-preserving purposes

Class IV: Evolutionary avatars

A representation of you that continues to learn, develop, change and experience – facilitated by AI.

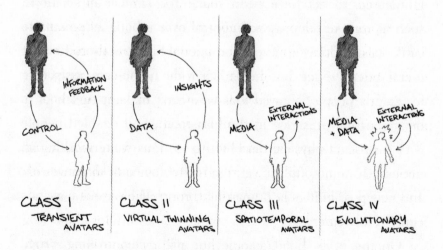

Figure 13: Four key avatar classes (AC-4)

CLASS I: TRANSIENT AVATARS

For these, think of the films *Avatar* or *Ready Player One*, which both illustrate the ideas of real and digital telepresence avatars. In *Avatar*, the main character, Jake Sully – a paraplegic former marine – is supplied by the military with a controllable biological body through which he can port his mind. The film is set in the mid-twenty-second century on a habitable moon known as Pandora, in the Alpha Centauri star system. Jake's larger avatar is modelled on a humanoid species indigenous to Pandora, the Na'vi.

Through telepresence technology you will be able to control your real-world avatar, which could be humanoid or android, may look like you or may not, eventually getting to the point where every movement and action you make is mirrored by your avatar counterpart – possibly with added or improved functionality compared to your biological body. Rudimentary versions have seen video calling to tablet PCs on a stick with wheels, allowing a person to be somewhat *present* in another place. Humanoid robots have been made for similar applications, such as remote telepresence control over robotic astronauts to carry out operations on the International Space Station. Robotic avatar telepresence has helped provide home-care assistance for elderly people, allowed kids who can't make it to school to have a virtual presence in the classroom, and enabled people with high-level physical disability to work as waiters. Although not exactly commonplace yet, the technology quickly moves on, and new possibilities in human-like, controllable robotic avatars consistently emerge.

On the other hand, people are already controlling *virtual avatars in games. Here, your avatar is a graphical representation of you, your alter ego or the character you're controlling in the game*. Through modern games, humans and virtual avatars are becoming more interconnected, allowing you *to move, run, jump, look around and interact with objects and other avatars in increasingly natural ways*. This is particularly the case when utilising extended-reality technologies, as in the VR depictions of *Ready Player One*. In the film, the main character controls his virtual avatar through natural actions with a range of external and wearable sensors, and even receives sensory feedback based on what *his avatar experiences in the OASIS virtual world*. Most times you go into VR, *you control a virtual avatar of some form, which can be made to look like you or not*. This is the interesting thing about avatars: they can be anything, look like anyone; they can be whatever you want them to be.

When it comes to creating copies of known people this class of avatar provides telepresence of the person, so they can be where they are not, acting through an avatar agent in the real or virtual world.

CLASS II: VIRTUAL TWINNING AVATARS

Avatars in this class are created as a digital, quantum (in the future) or other virtual representation of a person (or object or place, but let's stick to people here) and fed data from their real-person counterpart for the system to analyse things about them (as previously discussed with respect to AI and health). It is in essence a virtual copy of you. Your *virtual twin* (currently, in a digital age, this is known as a *digital twin*) lives in computers and exists not as a collation of atoms, but as a collation of ones and zeros – and when the technology is widely accessible, you may have a *quantum twin*, which will be a representation of you in a quantum computer.

These computers, still in their infancy, will be much more powerful than current machines. They harness quantum bits, or *qubits*, which rely on the phenomena in quantum mechanics known as superposition and entanglement to perform computation. A qubit is somewhat analogous to the digital bit in classical computing, only it can be in a quantum state of 1 or 0, or be in superposition of both states, essentially being both 1 and 0 at the same time. I can't do justice to the future implications of quantum computing here, so I'll leave that discussion for another time.

For Class II avatars, whether the computing is digital or quantum, it's all the same idea: you have a virtual twin of some form that exists as a Class II avatar, because it's consistently fed data about you. The first two classes both require the real person for control or data flow, but if the person is not present behind the real-time flow of data into the avatar for control or analysis, then this leads us to Class III. And now we hit the classes that can continue to thrive in usefulness beyond the death of their real-person counterpart.

CLASS III: SPATIOTEMPORAL AVATARS

These basically allow us to capture a person in space and time. They can be made interactive, essentially being as much as possible of the person's likeness – their looks, voice, gestures, mannerisms, personality, memories, interests – captured in an avatar representation with which we can interact in years to come. This is the space I find most intriguing and I quickly realised this was what I would have liked to explore had the relevant data from my grandpa been available (and similarly for my paternal grandparents). This can create beautiful, accurate and lifelike memories of an individual from the point in time the data was captured. In the realm of reality Class III avatars are most likely androids of people, like Android Phil previously mentioned. In the virtual realm these will be more like an advanced and interactive video recording, a new medium by which to learn about who a person was. Interactive playback of a virtual Class III avatar can be displayed in forms such as a video, giving a similar experience to a video call; VR, giving the feeling of *going into a memory and completely immersive experiences* (like those we'll discuss in the next three chapters); and AR, *bringing a perceived avatar hologram* into the real world – you could even sit them on your couch and chat (just don't expect them to drink any tea!).

CLASS IV: EVOLUTIONARY AVATARS

These take all of the Class I–III technologies to a whole new level and allow the avatar of an individual to continue learning. This form of avatar isn't reliant only upon data captured in time. Instead, AI allows the evolution of the entity to continue – drawing on new learnings, experiences and memories to simulate the person on which it was based, living on. But let's be real, this is very *Black Mirror* (a sci-fi series that depicts extremely dark and disturbing potentials of technological advancement).

In my opinion, ethical considerations go into overdrive when Class IV avatars are being created. Firstly, even if they are possible, I can't see too many good reasons for creating one. The avatar would evolve in a direction that is not necessarily, indeed is unlikely to be, the direction their human counterpart would have gone. With every choice we make, event we experience, interaction we have, we change – sometimes in a minor way, sometimes with large strides, but we do change. This means that from this very point in time you could go on to become one of an infinite number of possible versions of you, and this will remain true every time you read this sentence again. Internal thoughts, external stimuli, life events and a whirlwind of other *stuff* contribute to who you are and who you will become. Thus, creating a Class IV avatar of anyone seems a bit void of purpose, because it will never be a true representation of who that person would have gone on to become . . . only a single possible outcome from an infinite set of possible outcomes. So this is not something I choose to explore. That's not to say we won't see it happen. We likely will. But my happy realm is with virtual Class III spatiotemporal avatars and discovering the beauty within exploring those 'captures' of humans in space and time. And what an exciting realm it is.

What is *you*?
In life, it's who you are,
how you treat others,
what you're interested in,
what you think and what you do.
Beyond life, it's the family you've had,
the impact you've contributed,
how you've been documented,
and ultimately, how you're *remembered*.

22

THE LEGACY OF YOU

CREATING VIRTUAL JORDAN

Your very own Class II virtual twin has been created. Now what? Well, you connect all the data about yourself to it, starting with age, background, values and personality profiles and you get your genome sequenced. Then you continue to feed it ongoing data about your life. The closer your virtual twin can *mimic* you, the more accurate it will be as a representation of you, and this means it needs a lot of data to feed off. This will come from those health trackers and sensors you're wearing or may have implanted (which will become increasingly common in this new era), your social media details, where you go, what you eat, the exercise you take, your sleep patterns, even the people you hang out with. Everything has an effect on your health in some form, so data from everything will help your virtual twin better represent you.

Now here's where things get interesting. Since your virtual avatar can exist in a computer program, this means the program can be turned into a computer game, like *The Sims*, except as a simulation of real-world experience. This computer game could be made *to give negative reinforcement, like punishments in the form of losing life, having health points docked or being diagnosed with an illness. It could also be made to give positive reinforcement, as many games do, with rewards giving longer life or extra health points, doing more of the things you love and discovering new quests and achievements.*

From a simple structure of a game, we introduce machine learning AI into the mix. In particular, reinforcement learning that's great at learning to play games. First, the system *compares your virtual twin avatar to that of many other people and recognises patterns in the data.* It then provides you with some interesting insights into what you might be susceptible to, the potential for allergies or illness and even if you might have a predisposition to something more serious, such as cancer or diabetes. Second, the AI could start to play this game as it would a game of Go. *It can take control of where the character (i.e. your virtual twin) goes, what it eats, its cash flow, the exercise it takes and much more. Playing out many pathways of your life* could allow the AI to give you plans and powerful insights into what might keep you healthier for longer, and what could allow you to lead a longer life of increased quality. It won't give you all the answers, nor will the answers it does give you always be 100 per cent correct, but it may be able to *play out seemingly infinite pathways* to help give you a higher probability of a happier, healthier life.

Would you want this fairly extreme scenario? Or would you prefer to leave it up to *chance*? Furthermore, knowing about something doesn't mean we do anything about it. Knowing that certain things such as junk food or alcohol aren't good for our bodies doesn't stop us possibly liking these things anyway. If it's to be useful, an AI approach to maximising our quality of life may need to include our individual guilty pleasures – after all, we are *human*.

This virtual twinning concept also works on smart devices, such as production-line robots, where data can be continually transferred to the twin and analysed. It can work on pretty much anything with connected data-gathering methods, even buildings or locations – or, more audaciously, entire planets. With the rise in the internet of things – where household devices, electronic goods and a range of emerging sensors collect data about, well, everything – virtual twinning will become increasingly useful and will be utilised to automatically monitor the lifespan and required maintenance of these devices, and the ecosystem they form will allow insights across the interconnected systems so they can be optimised by AI for, say, energy efficiency and preservation.

For now, all we need to know is that people, objects and things, locations, environments, countries, planets, galaxies and maybe even universes, can at various points in time be constructed to have virtual twins, so that we can play out future pathways like a video game through the convergence of our advancing technologies. This pairing of virtual and physical worlds provides analysis, monitoring and predictions – to prevent problems before they occur, optimise systems and simulate future possibilities.

It seems that many governments can also benefit from AI virtual twin advisory systems.

Your relationship with avatars could even go beyond your Class II virtual twin. Have you ever created messages for your future self? Or created videos to share part of your life with others? These are just a few areas that can be taken into a whole new realm with Class III spatiotemporal avatars. Imagine having *even a brief conversation with your younger self as you were when the avatar was created*. Most of us look back and think how childish or naïve we once were, before we came to gather all the *wisdom* we have now. Sometimes it's also good to check in with our core values, many of which we've built from a young age. Are you making your younger self proud? What if your own avatar were the one to check in on how you're progressing with those big dreams you once had? We all keep items that create a sense of nostalgia, look at photos and videos, listen to songs we grew up with and from time to time we reminisce about who we once were. Getting to interact and actually converse with a Class III avatar of our younger self takes that to a whole new level.

Throughout my life, I've made mental notes for the future, on everything from clothing style (which I later renegotiated with my inner fashion critic) to the skill pathways I wanted to focus on in life. When I was leaving uni, I had my iPod engraved with a message to remind myself in the years ahead what I'd hope to keep as a core ambition in my life, knowing that I would look at it frequently while listening to music.

This is the phrase that reminds me of my purpose-fuelled ambitions from wanting to find a way to positively affect one life, even in a small way, to wanting to improve as many lives as possible. With a limit on the number of characters allowed, I had to choose my words carefully, so I made each one count:

One Life.

Persist to Improve Many.

Now what about having these various avatars of yourself, taken at different points in your life, remain around for future generations of your family? I find the idea of immortalising ourselves and our loved ones throughout our lives, leaving a legacy for future generations, an exciting one. In a step along the path towards discovering what creating avatars could be like, I was an early investor in 2015 in the first company I came across that seemed to be working even remotely in this space – Humense. This startup team based in Sydney was working on *photogrammetry* (creating 3D models of people from a range of 2D photos) and *volumetric* (lifelike 3D recording) video captures of people.

This encounter led to us creating a virtual avatar for my 2016 TEDxSydney talk at the Sydney Opera House. It starts through a chat with the founder Scott O'Brien and the team about the possibility of me meeting a virtual avatar. But the team at Humense are finishing up a massive project, so I just can't see the timing working out. *Becoming Superhuman* has finished filming and I've been hard at work with my Psykinetic team on a few other projects. Ten days out from the big TEDxSydney day, I go into the early rehearsals for feedback from the organisers, with nothing more than five rough ideas. After I present these briefly, the head organiser, Edwina Throsby, says she's interested in the avatar talk.

'Okay, the problem with that is we haven't made it yet,' I inform her.

'That's fine. I know you'll get it done,' she quickly shoots back in a manner that's comforting, firm and trusting all at once.

'Yeah, that's all I needed to hear. If you're happy for us to have a shot at it we'll give it our best.' I need to get cracking on this ambitious project with its difficult deadline.

So I'm going to meet my virtual self on stage. I've never seen anything like this and I have no idea how it might play out. The first thing to do is check in with the Humense team, who have now just completed their big project, and get them on side. We make a game plan of our requirements and the help we're going to need. We start hyping up our energy in preparation for a challenging time ahead. *Let's do this!*

Humense already have a camera rig constructed. It looks like a circular tent that glows bright with white light when turned on. Actually, it looks like a time machine. The walls and roof are lined with supporting poles, lights and 84 cameras that face the centre of the space. When you step into the centre, you feel as if you're about to be teleported to the other end of the galaxy. It's awesome.

I get stuck into the vision of how I want to present this feat on stage. Our eclectic team is made up of Jacob, a technically brilliant guy who's managing the team; Amber, an amazing force of a human behind the business development alongside Scott; Victor, a master in 3D computer graphics; and Ben, a young, randomly tattooed, bearded hipster with big glasses and a cool mix of genius computing and programming skills. Ben is the newest member, having joined the team the first time we set up a camera rig prototype. We were in a large room in an actors' studio to which we had temporary access, and I was standing near the door chatting with the team, who were all sitting on chairs in a circle. I noticed everyone look to my right and naturally my head turned that way too.

'Whoa!' I yelled and stepped back. This hipster dude was standing right next to me just staring at the camera rig through literally the thickest glasses I've ever seen.

'What you guys doing?' he calmly said, raising a giant water bottle and drinking through a straw, making a nice slurping noise. No one said anything, we just looked at him.

I broke the silence with, 'We're . . . uh . . . making a volumetric capture rig.'

He didn't turn a hair. 'Oh cool. Hey, I'm Ben.'

After introductions, Jacob asked what he was doing there.

'I do all the AV stuff for the actors' studio. But it's pretty boring. Not much is happening around here today . . .' He looked around the room again, and although his face didn't change at all, I could see the interest in his eyes. He slurped his water and asked, 'Need help with anything?'

Jacob leapt on the opportunity: 'Really? Know how to set up operating systems and get these old laptops going?'

'Sure,' Ben said, then walked over to the table with Laptop Mountain, sat down, slurped his water again and started working.

A bit stunned for a moment, we let him go for it and then we divvied up the remaining tasks. Ben stayed all day and turned up the next morning too. Long story short, we kept him. Sometimes this is how you meet the best members of your teams or the best people in your life – sheer strange randomness.

Our plan for TEDxSydney includes some very long days and nights, a rough idea of how it will play out with a little contingency time for unforeseen challenges ahead. First up, a few new computers are set up with a stack of computing and graphics power built in. I head over to the tent-from-the-future camera rig. Lights on. Cameras ready. I have a short one-minute script I've memorised of a few responses and interactions I want my avatar to exhibit in this on-stage demonstration. The minute is up and I get there with the script. Now to try to turn this camera footage, all 84 cameras' worth of it, into a volumetric 3D avatar.

As always, we're hitting one problem after another and need to figure out how to conquer each hurdle as it appears. The software works by using what are called *landmarks* in the photos (i.e. noticeable features such as the eyes) to figure out which cameras can see those landmarks, and then extrapolate the position of those cameras in 3D space. This is similar to the way you can figure out roughly where you are in your lounge room based on the objects you can see and the angle you're viewing them from. Once the system knows where the cameras are, it can reverse the process, and in a similar way to stereoscopic vision, the pixels of the subject in the image (that's me) are pushed into 3D space to make an avatar. Well, that's the idea anyway. We're quickly realising we're going to have trouble processing all this heavy data into a 3D avatar.

Time to call on our friends. Nick from my Psykinetic team joins to help us get over the line, and my brother Alex and a few other friends also become available to help out. What's needed right now is processing and graphics power, and a lot of it. After social media callouts and a bunch of phone calls we have mates turn up with their gaming computers. Jacob gets to work networking them into a beast of a computer cluster. But when the estimates arrive from the machine as to how long it will take, we see it requires five days to process the data. We have merely two days left!

Hipster Ben is on the case. He has been looking into accessing cloud computing so he'll give this a go. The process takes him a whole night but he gets there, accessing supercomputing power through a few different cloud services to chew through a now smaller version of the processing. He says even with this it's going to take thirteen hours. The time is 8.30 am and I'm on stage tomorrow morning at 9 am. We head into the Opera House for set-ups and testing, despite not yet having our complete demo to display. What we do have is a VR version of the stage Victor has put together. With all the VR set-up problems being smoothed out on the main Opera House stage, we plan out how we'll make this demo part run smoothly tomorrow morning.

Back at the workspace, moving in and out of the virtual world starts to get confusing and we find ourselves creating terminology to aid in the discussions to make it a little easier for us to all get into the same headspace. Our Class III avatar, the one filmed volumetrically from the large camera rig that will be *independently going through its pre-recorded movements, is aptly named Virtual Jordan*. We talk about objects like cameras between the real and virtual worlds and are starting to tie ourselves in knots if any one of the team isn't keeping up with conversation. For the same sort of reason I had to create *virtualitalics* during the writing of this book.

Time is moving fast. It's 7 pm and only an hour until the TV premiere of *Becoming Superhuman*. I had planned with Mum to take the team over to watch at her place. It's also only a few hours until the render of Virtual Jordan completes and will be ready for us to import into our VR stage. I don't think now is a good time to break focus. Much work is yet to be done tonight. So I call Mum and let her know we're not going to come round or watch the premiere, to which she says in a tone that sounds quite crushed, 'But . . . we have champagne. You can't miss your premiere.' Immediately convinced, I grab the team and we head over to my parents' house.

The half-hour show begins just as we arrive. I quickly realise how nervous I feel. Will it enjoy the response we have hoped for or will it fall flat? One of the key drivers behind the documentary was to change perceptions, hopefully globally, about disability and inclusivity, human determination and endeavour, and how technology is opening up new possibilities towards our super-human future. I wanted the world to get to know Riley for who he really is – a brilliant, driven, compassionate individual who is capable of achieving anything he sets his mind to. Mindset is everything. With our collaboration throughout this documentary we intended to take people along on our adventure to spread this sentiment far and wide. Thankfully, it captured everything we aimed for and more. Amazing responses start flowing in through messages and social media. We briefly celebrate, have a very happy phone call with Riley and his family, and it's time to get back to it.

When we get back to the workspace again, the time has just ticked over to 10 pm. Our render is complete, so we go to import it into our game-development engine to prepare it for the Vive VR headset. But it seems the import might take a while to process. As we watch the system calculate the time, a new wave of nerves come about. It estimates nearly four hours to import.

I'm starting to think tonight isn't a night for sleep. We continue to fine-tune everything else, so that when the import is complete we'll just *drop Virtual Jordan into the scene and it should be ready to go*. But when that time comes, at nearly 2 am, we're given a new surprise we weren't expecting. The visual looks good but the audio is not so good. It's lagging behind the video, meaning that *when Virtual Jordan speaks, the audio doesn't match his mouth movements. It is actually pretty creepy.* So this must be solved. A few more hours pass while I gather the content and put together the slides for my talk, working through what I want to say in my short eighteen minutes on stage. The time is now 4.30 am and *the audio issue is no better than it was before, because it varies as to how badly it lags* and so there isn't a simple fix. Executive decision time: *Virtual Jordan won't say anything;* I'll do all the talking. *The filmed snippets we choose to make him up are those where he isn't speaking, meaning he just stands there breathing and looking around. Silently. Lifelike.*

Good enough. Hopefully we can pull this big talk off in the morning. For now, it's time for a few hours sleep.

The alarm set for 6.30 am comes around in the blink of an eye. The crew assembles. We head to the Sydney Opera House. I'm quickly realising I feel terrible with a bit of a rough throat.

Setting up, it fast becomes apparent the program of the VR avatar demo that I'll be showcasing during the talk is being temperamental. Sometimes the program doesn't launch and the system needs to be reset. Unfortunately, the computer controlling it offstage, which Jacob is running, shows it working even when the headset is streaming out nothing, the big stage screen showing up blank too. Each time this happens, the computer is reset and we relaunch the program. It works. Then it doesn't. Then it works. And it doesn't again.

We've run out of time. The next time it runs we make the decision to leave it there and hope for the best. I'm now starting to cough and my throat feels like it catches whenever I talk, making it worse. An assistant finds me cough lollies and tea as I listen to the opening addresses and the first talk. I'm on second.

Edwina comes over to say hi and wish me luck. 'I'm so excited to see what you've all put together. It's all going to work okay?' she asks.

'Honestly, I'm about 40 per cent confident on that one,' I say through a raft of coughing, which is quickly taking over. I'm not entirely sure why 40 per cent confidence comes to mind, but I feel pretty solid about it. 'It'll be fine, though,' I reassure her. 'I'll figure it out.'

Now I sit back and focus on calming myself while throwing back lozenges and slamming down the tea. Coughing uncontrollably into the microphone and deafening a crowd of a few thousand at the Sydney Opera House is not how I envisaged this TEDxSydney talk playing out.

The talk before mine concludes. The crowd applaud the speaker as she leaves the stage. Edwina walks out to thank her and introduce me. My mind is becoming pretty clear. I'm not thinking about it all and I'm not stressing. Whatever happens,

happens. *Just get out there and give it your best*, my inner voice says. Edwina invites me to the stage. I walk out to a large, vibrant crowd that's applauding and cheering. *All right, let's do this.*

The second I begin speaking, I realise my first few words come out without coughing. Good start. Now I'm piecing together my talk on stage as I go. Although it isn't easy to speak, I'm getting through it just fine. I briefly introduce the ideas of living somewhere along the continuum between real and virtual worlds, and my thoughts on avatars at this point in time. I talk about my grandfather and how an avatar could possibly work. But I haven't yet segmented these ideas into classes and so I describe something that now seems to fall somewhere between a Class III and a Class IV avatar. The entire time I'm speaking, a bit of lingering doubt is hanging there in the back of my mind. *Will this demo work?*

I get to the point in the talk where I'm going to introduce the demo we have so far, of our first 3D volumetric video capture of me as an avatar in VR. I walk over, talking about the collaboration between my Psykinetic team and Humense, as a video montage illustrates the work that has gone into this over the past week. I have a growing feeling of uncertainty as I take each step over to where the VR set-up is located. Nick hands me the headset and controllers. I close my eyes, slip the headset over my face and hold my breath for a moment. Upon opening my eyes, *the first thing I see is the glowing virtual stage without a roof, and a starry night sky above. It works!!* But can the audience see what I'm seeing?

'Now I'll wait till you guys can see what I can see,' I say, *referring to the video feed of my virtual perspective* on the

projector screens. *'Let me know when you can.'* And they do, excellent!

'Now how would you respond . . . think about it for a moment . . . how would you respond if you found yourself face to face with . . . yourself?' I turn to my left just in time to see my avatar turn from facing the crowd's direction to looking at me. I instantly feel a shockwave flow down my spine as I hear loud, rather nervous laughter from the audience. *The immersive factor makes my mind tell me this avatar is real. He looks like me, but he's not me. It's a weird feeling coursing through my entire body as I start to walk around my Virtual Jordan avatar. He holds a* 3D real-world space, so I should be able to walk around him physically. I feel a pull on the cable connected to the headset I'm wearing. It's stuck. I casually head back where I came from. I grab the cable and give it a tug. It's hooked on something on the stage and won't give me any more slack. *So I get the best view of Virtual Jordan I can, before briefly asking the audience to imagine the empathetic possibilities if you could step into the shoes of someone else and see through their eyes.* I finish the talk, relieved the technology worked. Hugs all round for my team when I leave the stage.

The next time I see the team I tell them I noticed when I was on stage that *Virtual Jordan's right shoulder blade was lower than his left,* but Jacob assures me the render came out fine. I've been told for a number of years that my shoulders aren't in alignment, and my right shoulder and shoulder blade are dropped compared to my left, but I've never seen this myself. When we look in the mirror we often don't see ourselves the way others do. Our self-image can sometimes make us notice certain things about ourselves while other things can go completely unnoticed. *This external, objective experience of viewing myself* has unlocked a vision I've never really been able to see.

This leads me to visit a podiatrist, who tells me my legs have a slight misalignment (apparently not uncommon), which means misaligned hips and shoulder blade and dropped shoulder. It really emanates upwards. I have orthotic inserts for my shoes custom-made to prop my right leg up, so now I'm all good and in alignment. It's an interesting side effect of viewing a VR avatar and illustrates the sort of insight that can be unlocked with the power of a technology like VR. It unleashes the imagination. It helps us build seemingly impossible ideas. It gives us an unprecedented suite of new tools with which to create and discover.

Technology is never meant to replace humans,
but if we're creative and thoughtful with it,
we bring to life new positive forms of human experience,
new forms of human memories,
and new forms of *human*.

23

THE UNCANNY VALLEY

YES, VIRTUAL YOU MAY BE EVEN CREEPIER THAN THE REAL YOU

It's October 2017. I'm at a technology conference in Adelaide and finding myself staring a little while having a very casual conversation with Japanese engineering professor Hiroshi Ishiguro about his research. I ask him questions to keep him talking while I'm thinking about this intriguing yet somewhat eerie situation. See, just a few minutes ago I was standing in front of one of his androids, made in the image of him. A real android Class III avatar. It looked strikingly close to him too. Granted, the real professor doesn't quite look 100 per cent human himself – with cosmetic surgery helping him meet his creation somewhere in the middle.

The android version, which speaks with the professor's voice, is constructed using silicone rubber, pneumatic actuators and electronics, and features hair directly from the head of its creator. When not moving it can be quite deceiving. When the android does move to speak, my mind immediately picks up on the small details that aren't quite right – the jagged head-turns compared to those of a human and the lack of human nuances like microfacial movements. The sheer engineering prowess that has gone into creating this robot should be applauded, and the convincing nature of it gives us a glimpse of a sci-fi *Blade Runner* Replicant-style future. Professor Ishiguro is a very interesting man who raises deeply philosophical and important conversations with his work. But why is it that we can find an android robot – one that in many respects looks deceptively human – well, a bit creepy?

It's due to a largely universal phenomenon in aesthetics known as the *uncanny valley*, which describes the relationship between human-likeness of an entity and our natural affinity (or not) for it. This helps us understand our response to a robot, animated character or avatar, depending where along the human-likeness scale they sit. It's not an exact science but more an observable phenomenon. As an entity becomes more human-like, with such characteristics as mannerisms, voice and personality, we have a natural affinity for them. We like them more. But only up to a certain point. When the entity becomes *too* human, it's common that our affinity plummets, our threat detectors go off and we can feel a little freaked out. It seems too *creepy*. This is the uncanny valley. The only way back out of the valley while moving in this direction of human-likeness is for the entity to be so human-like as to be practically indistinguishable from a real human. When we get to this point, people's responses can be quite positive, even producing mirror neuron activity as if in the presence of another person.

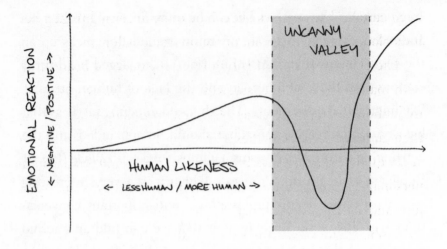

Figure 14: The uncanny valley

My Virtual Jordan avatar could have been pretty creepy were it not for the realism, maybe only just emerging from the valley – though due to many stray pixels not quite finding their precise place in 3D space, he does look a little like he's caught in a dust storm. Many animated movies have overcome the uncanny valley and learnt through experience. *Shrek* was the first computer-animated film to feature human characters in lead roles, so there were many unknown challenges in its creation. Shrek is an ogre character with numerous funny human qualities. He's really quite likeable. In the first version of the film, however, his love interest, Princess Fiona, was created to be a beautiful and realistic human, with an array of advanced techniques (for the time) going into her realism. But she went too far along the curve in human likeness, becoming too real at times, yet not real enough, and plummeting down the valley. The disturbing nature of her 'hyperrealism' caused some children to cry when the film was previewed to test audiences, so the filmmakers reanimated her so that she was more cartoon-like and less like a human simulation, bringing her back along the curve and out of the uncanny valley.

The characters in *WALL-E* cracked the concept beautifully, with many of the little characters showing so much personality without being in any way mistakenly human – they owned their animated robot bodies but their likeability came from such qualities as WALL-E's naïve, curious and empathetic personality and gestures; the love that blossoms between him and EVE; and the hilarious levels of frustration in M-O (Microbe-Obliterator), the small cleaning robot, every time he sees contaminants around.

In my show *Meet the Avatars*, I speak to Mike Seymour, an associate lecturer in the Business School at the University of Sydney, who also used to work in film digital FX. We talk about the uncanny valley and the idea that it may be possible for it to be 'flooded', so that it's less about the appearance and more about the emotion. In other words, if we get to that part where a non-human entity unnerves people and falls down the valley, we can perhaps lift it out again by improving its emotions and human-like interaction. From what I've seen to date, emotional design plays a massive role, but the visual appearance and movement of created characters still elicit some of our strongest gut reactions.

To see more about this, I travel over to the University of Southern California in the US, where they experiment with a range of these ideas. First up, I get to sit with an avatar psychiatrist. She has been computer-generated rather than recorded from a real person, so she definitely looks animated. The persistent presence and seeming eye contact as she looks at me from the screen in front of me is surprisingly engaging. After my first few interactions she gets a few things jumbled up and doesn't quite understand my responses. Maybe it's my Aussie accent. It does break the engagement level a little.

The one I've been looking forward to is next. I head over to meet a Class III avatar of a real-life Holocaust survivor named Pinchas. I've heard that there's a holographic display version of Pinchas to interact with, but walking into the room, all I can see is a large flat-screen TV turned sideways into portrait mode, with an idle repeating video of Pinchas sitting in a chair, just looking forward at me. Honestly, I was so looking forward to the holographic version that I'm taken aback. Surely this won't give the feeling of Pinchas' presence when I'm standing here looking at a 2D screen. Of course, I still give it a go, but my expectations of the experience have lowered.

Pinchas Gutter was born in 1932 in Łódź, Poland. He has since been an educator and guest lecturer on the Holocaust. This project was set up as a collaboration between Heather Maio of Conscience Display at the University of Southern California's Institute for Creative Technology and the university's Shoah Foundation (established by Steven Spielberg). Pinchas was interviewed while sitting in a large high-tech dome (which I also got inside during this visit), where he was lit by 6000 LED lights and captured from 52 surrounding cameras for a total of 25 hours of recording over five days – answering more than 900 questions thrown his way. This all forms video that plays continuously, triggered by verbal interaction, and moving back to an idle state of Pinchas sitting and listening in between.

Before I know it, something magical happens. I pick up the microphone and say hello to Virtual Pinchas, to which he responds with: 'Hello, I believe you have a question for me.' I'm already stumped, because I don't know much about his past other than that he is a Holocaust survivor.

'Tell me about yourself,' I begin with.

He proceeds to tell me a bit about his upbringing, where he was born and raised, mentioning his family with so much warmth. After telling me about a few more events, he mentions that he is a survivor of five concentration camps during the Holocaust.

'Five?' I didn't know this. Already his story sounds unique. 'Uh, tell me about the first concentration camp.'

Virtual Pinchas proceeds to tell me that when he was taken there with his family as a young man, the first thing that happened was his mother and sister were taken away from him. He never saw them again. He says that it is a very painful memory and mentions more events that occurred. Already it feels like my throat has caught. My mind is on the loss of his mother and sister. I want to go back and hear more about them.

'Tell me about your mother,' I say, to which he responds by describing her and how lovely she was. How he misses her. Already I'm getting lost in his story and I feel his presence, because in this interaction I am directing the conversation. That's exactly what it feels like – a conversation.

I have quickly let go of the fact I'm speaking to a sideways TV screen. Rather, I am now immersed in getting to know this kind human who has been to hell and back. He is exposing vulnerability and sharing with me his life story and what he has been through. That is the power of immersion and the feeling of presence – the human side of technology.

He has just taken me on a journey through some terribly painful moments in his life. Yet he's still so positive and warm. I now just want to give him a hug. This is definitely an effective mode of interacting with a person and getting to know them. It is confirming what I believed to be true, with a lower-tech solution than I thought was required – that a Class III virtual avatar can become a beautiful experience of human connection facilitated by technology. It can transcend time and space, immortalising one's memory. It's true that if done well, virtualising a person can be a wonderful thing.

Our near future will progressively see the utilisation of avatar technology in many aspects of life – from preservation of the memories of people like Pinchas to one-on-one communication with friends, loved ones, teachers and inspirational leaders; through virtual office meetings, training and surgeries; all the way to virtual assistants and even companions. It's a space that will bring with it a huge range of ethical issues to work through, especially given the possible misuse of this technology. And this misuse has everything to do with the uncanny valley. If you get an avatar out of the valley to seem so human-like it's basically indistinguishable from the real thing, this could be used to bring actors into movies who have already passed away. There have already been many attempts in films, using computer-generated imagery (CGI) and other filming tricks, such as when Grand Moff Tarkin, commander of the Death Star, was brought back through CGI in *Star Wars: Rogue One* – since the original actor, Peter Cushing, died in 1994. This may not have been the most convincing form of use, but as this technology advances, it will appear seamless in films. But there is a flip side.

A similar method of applying a person's face to the body of another in video, known as *deepfakes*, could also be used to make a puppet of important people, such as world leaders, putting words in their mouth and creating fake video footage indistinguishable from real footage – something a friend Hao, who I met in this documentary, extensively researches. The approach utilises a deep learning type of AI known as generative adversarial networks to perform a range of analyses and processes (often involving one artificial neural network generating footage and attempting to trick another artificial neural network discriminating real from fake), ultimately producing results like overlaying footage of the subject's face on the footage of an actor. This is in most cases done without the consent of the subject. These counterfeit videos will be something we need to be increasingly wary of and governments must address early. We need to treat video with the caution with which we already approach photos.

Governments and regulatory bodies will increasingly require scientists and technologists to advise on the many capabilities emerging from today's advancements, and to keep up with those changes while shaping the direction they take. Reducing funding to science research is one of the greatest mistakes a nation can make. Uncertainty and change is on the rise in this new world. Propagating fear is not the solution. What allows us to thrive is sharing our humanity, positive visions, ideas and action so we can work towards a better world – one that fosters knowledge, love and connection across the globe. The need to recognise that with each new advancement there is always a positive opportunity is paramount. And that's what we must imagine, must search for, address, understand, share and propagate.

For avatar technology, then, I want to explore the ideas that can enhance our nostalgia and human connection, and immortalise the memory of loved ones in as authentic a manner as possible. Taking my inspiration from wanting to have a virtual avatar of my own grandfather, I aim to locate some other grandparents who are willing to help me uncover what these innovations have to offer. For this adventure, I meet a caring and compassionate Italian couple. Both are fit and healthy. They've been together for more than half a century and have raised a large, loving family around them. Their names are Maureen and Michele.

24

HUMANS IN THE CLOUD

MAKING INTERACTIVE AVATARS
OF REAL PEOPLE

Maureen and Michele (pronounced Mi-kel), the beautiful couple who will be my avatar trailblazers, first met just over half a century ago. Maureen tells me how she knew the moment she laid eyes on Michele that she'd marry him, a cabinetmaker fresh off the boat from Italy. There's a lot of love and history with this couple and the big family they've raised in Australia.

I start by talking them through what VR is and giving them a go at experiencing it to understand some of its capabilities. I then describe my idea of creating an interactive virtual avatar of Michele. It feels surreal talking about it. I'm not entirely sure myself how we'll do it yet or how convincing it'll be. Will we fall into the uncanny valley or will we bring it out the other end so it's indistinguishably real, while remaining authentic? What will be the emotional impact on the family? Will it even be a positive experience?

I have many questions and minor worries whirling around in my head. But again, if we don't explore, our ideas will remain unrealised. Something I'm becoming increasingly curious about is the human reaction to a virtual avatar. I'm wondering if a person can exhibit mirror neuron activity, or get the feeling a person is present, if it's someone they know really well, as will be the case here with Michele – no one knows him better than Maureen – or whether this makes it harder. Will this have the same level of effect as I felt meeting myself, or the avatar of a person I've never actually met, like the virtual Pinchas interaction? The challenge here won't just be to create a Class III spatiotemporal avatar of Michele. It'll be to make it so seamless that Maureen feels she is in the presence of her real husband.

A number of factors will contribute to this desired response if done well. If the brain is completely convinced of the interaction, it may show a response similar to an in-person conversation. This is why face-to-face human interaction is so important. It is said that eyes are the window to the soul, and this seems to be particularly the case when we hold someone's gaze. Our brain activity can synchronise in some ways, our subconscious kicks into gear (picking up on emotion and empathy, social cues and micro-expressions), our brains converse and, seemingly, our souls connect.

This is the sort of magic I'm curious about when it comes to a face-to-face interaction with the avatar of a loved one. If it's real enough, it's possible our response to the avatar may be similar, or even identical, to when we are in their actual presence. This is what I'm setting out to achieve with the creation of Virtual Michele.

Unlike the previous design using Humense technology with a huge camera rig set up, masses of data and cloud computing to crunch that data for my avatar, this time around we need to do it quickly and low-tech in comparison. The focus here will be more on the interactivity component – to design a Class III avatar where the user feels the presence of Michele. Based on the timing of this documentary, my team from TEDxSydney is unable to reunite for this project. Our hunt for a new small team has located a couple of innovation specialists from the University of Newcastle – Craig and Gaute. I talk them through what we're aiming to achieve and we get stuck in.

With the use of a single green screen, a couple of lights and a single camera, we sit Michele down and prepare him to be interviewed for our data collection. Sitting behind the camera for good eye line, I get Michele to imagine he's speaking with his wife, Maureen. Once he seems comfortable enough we begin. I ask him about where he grew up, how he and Maureen met, other memories he wants to describe and any messages he would like to leave for her. We then start again for his kids, and again for his grandkids, and again for people he doesn't know. Why? Because we respond differently to family, friends and people we don't know. So I wanted the capture to work for whoever would be interacting with Virtual Michele in the future. We cover a lot of ground over three hours, giving us plenty of footage to work with for this project.

To build this Class III avatar, we start with a computer-designed *3D representation of Michele's head, and start wrapping all of our 2D footage around it.* A few extra steps make it easier to complete the avatar within the week and a half we have until the documentary's final day of filming. The team *builds a static body for Virtual Michele. The table he's sitting at, and the surrounding home environment, contain many bits and pieces* captured from their actual home, particularly photo frames, furniture and paintings. We utilise cloud services to make *Virtual Michele interactive through AI* similar to what you'd find in smartphone or home virtual assistants.

When you ask Virtual Michele a question the audio captured by the microphones goes straight to a cloud service, which puts them through a natural language processor that separates your words, and pulls out the keywords to gain context to the question as best it can. Through chatbot-style AI, the question is compared with the database of Michele's answers. If there's a match, *it plays the sequence of his answer.* If there's no match, *Virtual Michele just sits there and smiles and blinks. It's a little awkward for a moment, so you naturally move on to asking another question.*

The truth is, I feel as if *Virtual Michele might slip into the uncanny valley a little,* and I hope the simplicity of our design (compared with the TEDxSydney Virtual Jordan) will still be adequate. We make lots of manual changes on the computer, *adjusting Michele's recorded movements so they're more stable. We want to focus on the 3D feel, realism and smoothness. But if we do get him out of the valley,* how will Maureen and the family respond?

So many thoughts like these are flashing through my mind as we approach the reveal of our designs. With the documentary's last day of filming upon us, all this uncertainty is starting to come to light. Will the avatar be slightly scary or unnerving? Will it be a little underwhelming yet still give us insight into what this technology could achieve? Or will it be a positive experience, potentially even an emotional foray into the world of creating virtual avatars of loved ones?

I get straight to it, asking Maureen, 'Now, are you ready to see a virtual version of your husband?'

'It's all very confusing to me. I've been in the dark,' she says, and I realise she hasn't much of an idea what's about to happen. She hasn't seen the process, but I know we recorded some lovely words and messages to her from her husband.

'Let's just hope he's not better than the real one,' I kid.

'Oh my god, what's happened here?' Maureen calls out as she walks in to our set-up downstairs in their home, seeing Gaute in front of her with all our computer equipment, VR, cameras, green screen and lights. Now, after 50 years of marriage, she's about to meet her husband for the first time in VR. Having had a look myself I can't wait. I get Maureen to close her eyes as I put the headset on her. It's time for *the avatar to meet the family*.

Although my mind is a flurry of anticipation, the words find their way out of my mouth smoothly and calmly. 'When you're ready, open your eyes.'

'Oh my goodness! . . . Hello!' Maureen says as she opens her eyes, shocked to see a digital copy of her husband Michele sitting across a virtual marble table, smiling at her. 'Hello. Ciao,' he replies, switching

to their native Italian tongue as if he knows he's speaking to his wife. He follows this up with an even bigger smile.

'Castelnuovo, does that ring a bell to you?' Maureen says as if she's trying to ascertain whether Virtual Michele can recount any of real Michele's past memories. Good thing my interview sessions with him covered so much ground!

'Uh, it's a little village on the Adriatic side of Italy, in the south,' he replies.

She smiles a moment. 'Have you got anything that you would like to say to me?'

'What I like about Maureen, a big smile and uh, she's a really good wife.' You can instantly feel that he's not used to opening up and being quite so raw with his emotions like this.

'That's beautiful.'

'You know, I don't think I could have done any better,' he says with a cute embarrassed, loving laugh.

Before long, Maureen has started to immerse herself in the conversation. I can see she feels as if she's just sitting there with her husband. 'Can you feel his presence right now?'

I see her suddenly remember there are others in the real room with her. Calmly, and without turning away from Virtual Michele, she responds, 'I sure can. Just his blinking of the eyes. His smile. His . . . I just feel as though I'd like to put my hand out and touch him.' She extends her hand out towards his face. She can't reach, and nor would she be able to touch Virtual Michele's face in this particular design, so she opts for a warm wave instead.

This moment suddenly opens up a universe of possibilities. The team have done a fantastic job. Next up is Michele's daughter, Louise. She walks in, takes a seat and looks around, clearly not entirely sure what has been going on with our set-up.

'Don't really know what to expect here, do you?' I joke as I prepare the VR headset.

'No, I don't. I've got no idea.' She lets out a nervous laugh. I place the headset over her face and *she instantly looks around. 'I see my father and some of his artwork.' She turns to face him as he stares with a smile, waiting for a question. 'What are your childhood memories of your town?'*

His answer comes through in his Italian-accented, smoothly croaky voice, 'We played a . . . played a lot of football . . . but, with the home-made ball, you know. I had to borrow things from my mother to . . . stockings to stuff it up, with clothes to make a round ball. And we used to play, in the street.'

I can see throughout this chat that Louise is starting to tear up, understandably so.

When she removes the headset, she dabs her eyes. 'It's actually quite sad. I feel quite emotional actually, yeah. I think seeing my dad talk about his family is always an emotional thing for us, so . . . it's lovely.'

'Did it feel real?'

Still processing what she's just experienced, she replies, 'Yeah! Unbelievable. It's like he's sitting there . . . Waiting for him to pull out a coffee,' she says with a laugh before talking through some of the possibilities for her kids interacting with Virtual Michele in the future.

An incredible few outcomes here have convinced me that there's a wonderful opportunity for leaving a virtual legacy of loved ones. I'm also now sure it can provide a truly touching, human level of connection if done well.

It has been an emotional and eye-opening experience for us all. And for Michele, we have one more. Using smartphone mobile VR with Street View (which displays 360 degree filmed streets from many places around the world), we have prepared a surprise for Michele with the help of his family. I put the headset on him.

'Describe it to us,' I say.

'That's my front door there.'

'Haa, really?' I ask, delighted we have it right. *VR has taken him home, to the village he grew up in.*

'Yeah . . . Yeah.'

'How long has it been since you've seen this place?' I ask. But no response. *He is just looking around. A moment later he puts his hand on his head, looks down and starts sobbing.* I didn't realise his reaction was going to be so emotional.

'Want me to take this off?' and *he nods.* Without hesitation I quickly lift the headset off him.

'Sorry,' he says as he emerges.

'You're all right.' I'm really worried this has upset him. I know the place he grew up in means a lot to him. He speaks of it frequently, *and so does Virtual Michele for that matter.*

He turns towards me and speaks from the heart. 'That was beautiful.' A wave of relief flows through me. He loved it and that's what I had hoped for. I give him a big hug.

We've all gone through such an intense and amazing experience together. It's funny that I feel so connected with this beautiful Italian family, because, funnily enough, I was mixed up with an Italian kid at birth. His cot in the hospital I was born in was close to a window but he was sensitive to sunlight, so the nurses switched our cots to get him away from it. Next thing, Mum is showing my grandparents the wrong baby and they're telling her they don't think it is her baby. Years on, I'm the only one in my family who can't get enough pasta and Italian food, so my parents often joke that they got the 'Russo kid'. But now I have found this lovely Italian family to visit for great catch-ups and great pasta!

Having been through this particular adventure, I'm sure now that if the data had been captured of my grandfather, I would love to have a virtual avatar of him to interact with. I'm completely convinced virtual avatars of loved ones will find their way into our lives. We will need to be careful, of course, in how these are used, paying particular care with how individuals grieve. But with an approach sensitive to the intricacies of relationships and how we individually deal with death, Class III avatars of loved ones will undoubtedly bring about new forms of positive human connection.

In fact, while writing this section over the past few days, I had a dream about these concepts. In my dream, *a friend of mine had become a bit of a guru and she was asking me what projects I have on the go currently and which are some of the most important ones to prioritise right now. I talked about taking these avatar creation concepts further into an app. She asked if there was a need for it and I told her that I haven't been asked for anything more frequently than to make this technology accessible so others can create avatars of their loved ones too. She asked why this wasn't happening now. I stopped to think about my answer . . .* and then

I woke up. (Hmm, seems *virtualitalics* works for dreams too.) It's true, I've had people reach out about this technology frequently. So I'm making the decision now to begin the next phase and start working on the release of an app that can be used to capture memories in this form. If we can make this available, individuals can create avatars of their family and friends, aged-care workers can interview the people they care for to capture their stories and people can themselves choose to leave behind interactive messages for the future.

However we choose to create and adopt avatar technology, it will increasingly become commonplace in our lives. More and more humans, in one way or another, will transition into the cloud. We live out our lives in the biological constructs of our human bodies (at least for now), but beyond life we largely live through the memories of others – and of course in the data trail and avatars we leave behind. One day in the distant future, if you're not living an immortal, posthuman life through a Class I transient avatar or some other immortality advancement, and you've passed on from this life, there may be Class III spatiotemporal avatars of you having some great conversations and sharing your memory. So before that happens, ask yourself these questions: What sort of stories would you tell? What vision and values would you share? And what sort of legacies would you like to leave for future generations?

Fly this way . . . to the future.
Hope you're still wearing that cape!

CONCLUSION

OUR LUCKY SPECK OF STARDUST AND MOMENT IN TIME

What a journey it has been! Navigating the rapidly advancing world of science and technology has been full of adventures and there are many more to come. What drives me every day is an underlying vision for a better future, inspiration from human endeavour and a strong purpose to improve lives and fight for a better world.

Each day I think about today as if it's the past, looking at the world (as advanced as it is) as if it's a place I'm looking back on, the way people sometimes look back and try to remember life before the smartphone, before the internet or before the personal computer. This helps me project forward and think about what's coming next. I also increase my self-awareness and perspective by imagining myself at the end of my life looking back.

Did I work towards the things I believed in? Is there anything I regret not having done in life? Was I true to my values and integrity, and authentic to who I am? This helps me remain clear about my path, make decisions and prioritise where to use my most precious resource. That precious resource: TIME.

We know we have one life . . . well, at least one. Big changes from science and technology in the theorised transhuman and posthuman eras may change that. But for now, we know we have one life. You're lucky enough that what you do with your one life is up to you. I know, within me, that I want to use my one life to improve the lives of as many others as possible. I want my future self to know that I did my best to be a driving force behind positive human and technological evolution as we move into the future. That I worked towards harnessing these skills for improving the future of life and the Earth in any way I could. For my friends, my family, my future family yet to breathe existence. That I contributed to their world. And that I didn't mind thinking big. This has been the backbone behind so many ideas and ambitious undertakings. If you dream big, then take action, you'll get somewhere towards that goal. So set the bar high.

I return regularly to that engraving on the back of my iPod, to remind myself afresh of the ambition I've set myself for this life:

One Life.
Persist to Improve Many.

It has guided many tough decisions I've made since to follow my broader vision. Without this engraved message to my future self, I might not have had anywhere near the level of clarity as to which choices and directions would instinctively feel right. I've learnt that certain things are critical in achieving your ambitions: self-belief that you will do what it takes to bring about positive change; mustering up the determination and drive to keep at it; connecting and joining forces with a diversity of other minds; and reminding yourself of where you're headed so you don't lose sight of the path.

If we're living at the fastest rate of change the world has ever seen, and the slowest we may ever see again, then we must adapt and even find prosperity along with it. In this collaborative era of connectedness, we should have more power than ever before to steer that change. I've seen this in action. Movements for equality, for peace, for inclusivity, for unity. So let's all continue to treat this as an opportunity to have meaningful conversations and work together towards a world we can be proud of, for the future of us . . . the future of humanity.

From small everyday things like smiling more to brighten the days of those around you, to big existential shifts like contributing to solving climate change; to driving forward better energy solutions; to empowering communities in need; to helping our environment recover and thrive; and to improving life on Earth. You have the power to positively affect our world. No matter how small or how big, it all makes a difference. You make a difference. Technology is a tool, but as amazing as it is, it's not the tools themselves that will change the world, it's how we choose to use them. If we start with vision and purpose, our technological toolkit will help us bring our ideas to life, and even make the impossible possible.

This vision for a better world is needed now, more than ever before. Our world is facing many massive challenges as we overdraw on the Earth's finite resources, feel the ongoing effects of the environment fighting back and deal on and off with big hits to our world – from extreme weather patterns to massive societal shifts to pandemics. And the road ahead will be bumpy too. Now is the time to act. I'm optimistic that one day we'll look back and remember when we almost lost control and almost suffered en masse as a consequence. But we didn't. We came to our senses. We strived for a better future and found ways to turn it around. We acknowledged the beauty and necessity of our natural world, harnessed our advancements in science and technology and, driven by the power of people and our imaginations, we forged a new path.

See, competition among our own species may have helped us push the abilities, consciousness and collective intellect of humans to the advanced levels of today, but at this point in humanity, it is crucial we move into an era of collective *collaboration*. The threats we face as a species, as life on Earth, are massive. Much of this has been brought about by our persistent conquering of the globe. Now to avoid those threats becoming irreversible and disastrous to all life, we must connect, we must act and we must work together to ensure there even is a future, and furthermore a bright future. For our families, for our future generations. For life.

This is our world. Our speck of stardust. We're cosmically lucky to have come to be exactly where and when we are – inhabitants of this planet, spinning on its own axis while orbiting around the sun. Our entire solar system but a tiny dot in the universe. Our ideal temperature band at our distance from the sun, our atmosphere and our water have helped yield and harbour life. More than 4.5 billion years of the Earth's history, through incredible planetary changes, land movements, eras, evolution of various species, development of biological ecosystems and a vast raft of other factors, have led you to being right where you are right now, alive and reflecting on life, at this very moment in time.

And that, my friend, is a very, very special thing.

So, how do I feel about technology? Well, I believe that we cannot stop the rapid rate of change. It continues and only slows a little here and there when we face big global challenges. So instead we must embrace and shape it, and I choose to approach this through developments that are:

Humanity-, life-, nature- and planet-inspired.
Science- and technology-enabled.

If we rewrite our future into a better one, we don't let our fear from all those dystopic sci-fi stories rule us. Instead, we realise technology is an enabler that can help us achieve our dreams.

What strikes me as a common thread throughout these relentless advancements in science and technology is that we seem to be constantly learning more about what it means to be human. Whether we're sending humans and new rockets into space; building robots as companions to stave off loneliness; comparing the advantages of the human brain, creativity and empathy to the drawbacks of AI; creating bionic technology to improve and increase quality of life; conducting groundbreaking experiments into the use of tech like VR in spinal rehabilitation; fabricating extended-reality worlds for the mind to explore; observing android replicas of humans and trying to understand our own emotional response; or capturing memories of those we love into interactive avatars for future generations to know – we're consistently inventing and innovating new stepping stones and forging ahead in our understanding of ourselves.

Despite the incredible possibilities that emerge from evolving technologies, it's actually their *convergence* that provides unprecedented opportunities. Just like those that have formed the backbone of this book. A combination of these, with a touch of some of the next advancements to come, will one day converge towards something I call *sustainable immortality*, where humans can thrive with massively extended length and quality of life and, more importantly, humanity can 'live long and prosper' without burdening the earth.

But there are many more mind-bending things heading our way and I can't wait to get stuck into more adventures with you, continuing to uncover that intersection between technology and humanity. They don't come out of nowhere, though. Many people in the past have laid the foundations for this transformational era we're now living in . . .

Great minds, thinkers, tinkerers, creators, builders, doers,
those who dared to dream big and were bold enough to act.
They are the giants on whose shoulders we stand.

In the ages to come, we must not act as a plague. Instead, we shall thrive as a species, and raise up and protect other life on earth along with us. Our advancements in science and technology have allowed us to gain many levels of understanding and influence over our world, but next our endeavours will move into an even bigger place.

Will we tackle our fourth dimension – time? Will we find ways to truly create our own next steps in evolution? Will we build massive virtual worlds? Will we uncover the holy grails of energy? Will we save our planet? Port mind into machine? Discover what the consciousness is and where it's held? Conquer the quantum realm and open up entirely new portals of possibility? Will we exit our planet, live throughout the solar system, or even move past it to explore the depths of space and the universe at large? And through all of this, will we be better humans?

There are many, many more questions to ask and answers to search for, so stay curious. What is certain is, from this moment on, you can become one of an infinite number of possible versions of yourself (which is why a Class IV evolutionary avatar will never be able to predict who you will become). Experience shapes you, but your underlying values must be your guide. So it's up to you to decide, at any moment in time, if you'll rise to the challenge and contribute towards improving our world however you can. We are the writers of our own destiny.

And this journey isn't over. No, no, not even close, not at all. Our adventures have only just begun. Our future starts now, so let's stop dreaming of a better world . . . and make one.

Good thing you're still wearing that cape.
Because now, my beautifully brilliant friend,
it's time for us to fly!

ACKNOWLEDGEMENTS

Writing a book was a massive challenge – and was itself an adventure! To have arrived at this point in time, having just completed everything and getting ready to hand it over to the printer, I am reflecting on the amalgamation of *A Human's Guide to the Future* and feeling a deep sense of gratitude to numerous people.

To my amazing publishing team at Pan Macmillan, thank you for your support and guidance through this book, and for really working with me to produce something I am very proud of – including my publisher Ingrid Ohlsson who has been a wonderful guide, managing director Ross Gibb for believing in me, senior editor Danielle Walker who has made the constant back and forth a pleasure and is nothing short of incredible, and our brilliant publicist Charlotte Ree. And my structural and copy editor Nicola Young who helped me challenge my claims and think deeper. A big thank you to Angus Fontaine for being my initial spark and so instrumental in my authoring journey, and to Mark Abernethy for getting me going with it all.

To my family for their constant encouragement. A special thank you to my greatest supporter, Mum, for listening to the entire journey of this book every day, and for beautifully stepping outside your artistic comfort zone to create the majority of illustrations in this book with me. I just love them and this book has instantly become infinitely more special with your art and input. To Dad for inspiring my future-thinking mind, and to Tristan, Alexander and Zohara for always being there to talk to – best siblings I could have asked for. Grandma and Rob, you are always a guiding light in my life. My beautiful niece Elora, your future is a big reason behind much of this. Our family who have passed, here's to keeping your memory and legacy alive.

To my extraordinary media manager Michael Rivette, thank you for being you. I am just loving our adventures and they're still just beginning. To the force that is Nanette Moulton (and the wonderful Winston Broadbent), your guidance is always uplifting. You are all much, much more than media managers, you are my family. To my documentary teams for all the adventures. To Monique Peachman, Anne Jamieson and my entire Saxton Speakers family, thank you for being so amazing as well as managing much of my busy life alongside the book so seamlessly. Thanks to Helen Hanna and Vicky Bardas (and Rashid, Peter and Isaac) for always keeping me ready for camera and adventures.

To all my family, extended family, friends and Psykinetic family who have been here for me, thank you for understanding that I did need to leave the real world from time to time and disappear into the dimension that is *A Human's Guide to the Future*. Special thanks to: Stephanie Lim, Nick Temple, Allison Reid, Victor Limsila, Eloise Cleary, Jess Irwin, Riley Saban, David Lau, Tanya Brown, Ben Panetta, Sophie Parr, Chris O'Neill, Meg McGowan and Graham King, Stuart Saxby, Sandy Ludman, Emma Boyd, Aimee Kubo, Tianna Li, Alexx Cass, Suzi Jamil, Sarah Nally and Jessica Thompson. To everyone who has been here for me during this journey and to those who inspire me, it means the world. Thank you.

INDEX

Forbes, Alex 257
Ford, Henry 124
Fourth Industrial Revolution 7, 8

Gardner, Howard Earl 98
gene sequencing 193
genetics 9
genomics research 9
Germany, robotic workers 88
glaciers, predictive technology and
    139–41
  black carbon 141
Go (game) and AI 95–6, 120–1
GPS 1
graphical processing units (GPUs)
    161
Gurdon, Sir John 199
Gutter, Pinchas 304
  Virtual Pinchas 304–6

Halo: Combat Evolved 232
Hanson Robotics
  Android Phil 84, 85, 281
  Sophia 41
haptic vibrations 213
Harari, Yuval Noah 156
Hawking, Professor Stephen 57,
    98–9
head-mounted displays (HMD)
    225, 229
  Headsight 230
  Oculus Rift DK1 239
  Oculus Rift DK2 205
  Rift 235
  Telesphere Mask 230
health technology 137–8
hearing
  bionic ears 190, 191
  cochlear implants 155, 186, 191
  sensory bypass 190–1
  sensory stimulation 189–90

heart
  biology of 151–3
  pacemakers 153, 154, 156
heart-rate monitors 1
Heilig, Morton
  Sensorama 230
  Telesphere Mask 230
heuristic programming 102
Hinchley, Ben 289–90, 292
hologram technology 225, 229,
    235–6
  perceived avatar hologram 281
Horror Tribute 240–1
human–AI symbiosis 183, 193
human interaction
  avatars 303, 310
  face-to-face 310
  mirror neuron activity 310
  robots 46, 74, 91, 129
human life, quality of 198
humanising technology 10
humanoids 31
  first 34
  labourers 48–9
  movement 47–8
Humense 288
  Virtual Jordan 288–98

I, Robot 96
IBM
  Deep Blue chess-playing
      computer 104
  Watson 104
imagination 155, 226, 229, 240,
    260, 298
imagined technologies as reality 19
immortality
  immortal jellyfish 197–8
  sustainable 326
  technological 267–8

technological innovation
  convergence of 326
  impacts 124
TEDxSydney 2016 288–97
Temple, Nick 207, 211, 214–15,
  217, 291, 296
The Terminator 19, 96
Tesla, Nikola 33
Tezuka, Osamu
  Astro Boy 35
Third Industrial (Digital) Revolution
  8
Thomas, Nathan 240–41
thought-controlled intelligent
  machine (TIM) 113, 114, 159,
  166, 180
  brain–computer interface (BCI)
    113, 114, 173–5, 182
  field of vision 167
  prefrontal cortex filter 177
  trial participants 178, 179–80,
    180–1
Throsby, Edwina 288, 295–6
Tibet 4–5
  Tibetan glaciers 139–41
time, the fourth dimension 327
Time-Captain Studios 4
time-of-flight principle 168–70
  echolation 168–9
transient avatars (Class I) 277,
  278–9
transport, innovations in 124
Tron 230–1
Turing, Alan 102
Turing test 102, 134

uncanny valley 301–3, 306, 312
United States 4
University of Newcastle 311
University of Southern California
  303

Institute for Creative Technology
  304
Shoah Foundation 304
University of Technology, Sydney
  (UTS) 15, 20

Vietnam 4
Virtual Jordan 288–97, 312
Virtual Michele 311–17
Virtual Pinchas 304–6
virtual reality (VR) 9, 194, 224,
  225
  camera ball 234–5
  'empathy machine' label 266
  exposure therapy using 243–8
  face-to-face interaction 310–11
  fictional portrayals 230, 231, 232
  future possibilities 267
  gaming and 231, 263
  historical developments 230–1,
    235–7
  human connection, experience of
    305–6
  immersion, power of 235, 243,
    266, 305
  mindfulness, use for 265
  mobile VR with Street View 316
  needle injections, use in 254–5
  occupations, use in 250–1
  omnidirectional treadmill 232–3,
    262
  phantom limb pain 253–4
  portals to 229–31
  psychological effects 240,
  reality–virtuality continuum
    225–6, 231
  rehabilitation, use in 4, 251
  school students, projects
    developed by 256–9
  Tilt Brush 265
  virtual stage displays 221–3, 225